Stretch Your Thinking

ENRICHMENT WORKBOOK

Harcourt Brace & Company
Orlando • Atlanta • Austin • Boston • San Francisco • Chicago • Dallas • New York • Toronto • London
http://www.hbschool.com

Printed in the United States of America

ISBN 0-15-311079-1

6 7 8 9 10 085 2001

CONTENTS

CHAPTER 23 Organizing Data

CHAPTER 24 Making Graphs

CHAPTER 25 Facts to 18

CHAPTER 26 More About Facts to 18

CHAPTER 27 Multiply and Divide

CHAPTER 28 Two-Digit Addition and Subtraction

Name ______________________________

One-to-One Correspondence

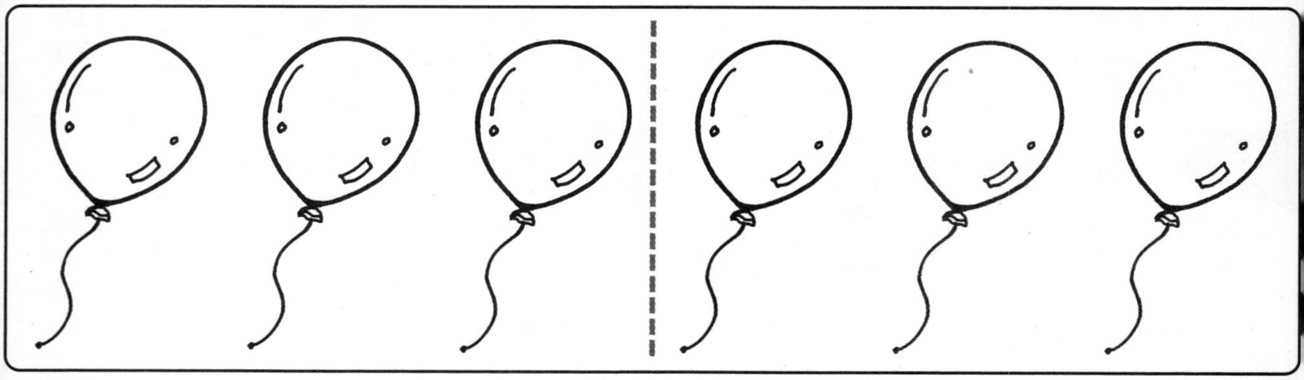

1.

2.

3.

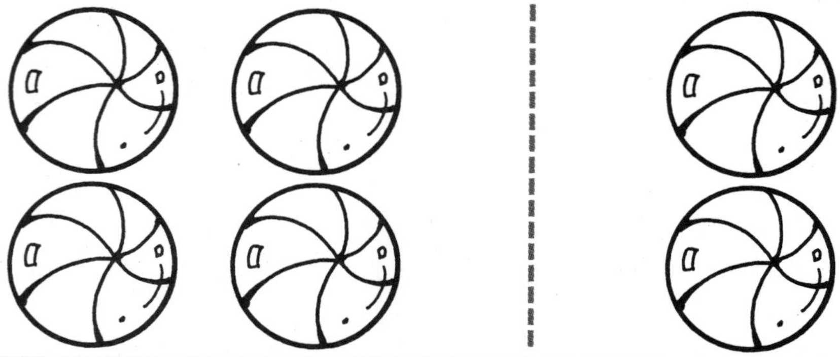

Make the group on the right look like the group on the left. Draw what is missing.

Name ______________________________

GETTING READY 2

More and Fewer

1.

2.

3.

4.

1.–2. Circle the fishbowl that has more fish.
3.–4. Circle the fishbowl that has fewer fish.

Name ____________________

Numbers Through 5

Color each bird blue.
Color each duck yellow.
Color each frog brown.

Color each tree green.
Color the bench red.
Write how many of each there are.

Name ____________________

GETTING READY 4

Numbers Through 9

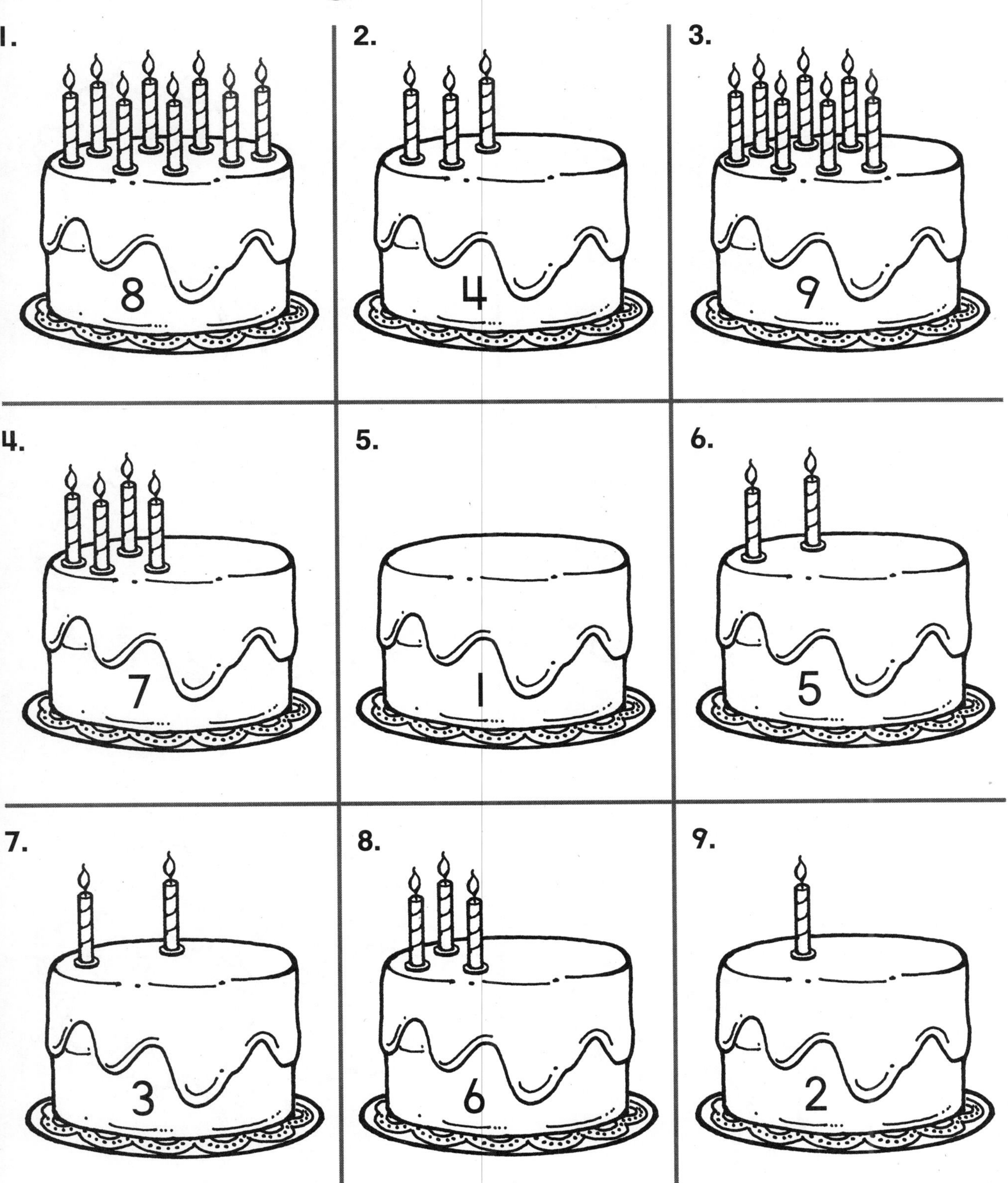

Look at the number on the cake. Draw the missing candles.

Name ______________________________

GETTING READY
5

Ten

1.

2.

3.

4.

5.

6.

Draw apples to make each tree have 10 apples.

Name ____________________

Greater Than

1.

2

6

2.

3.

4.

Count the animals. Write the number.
Circle the number that is greater.

Name ______________________________

Less Than

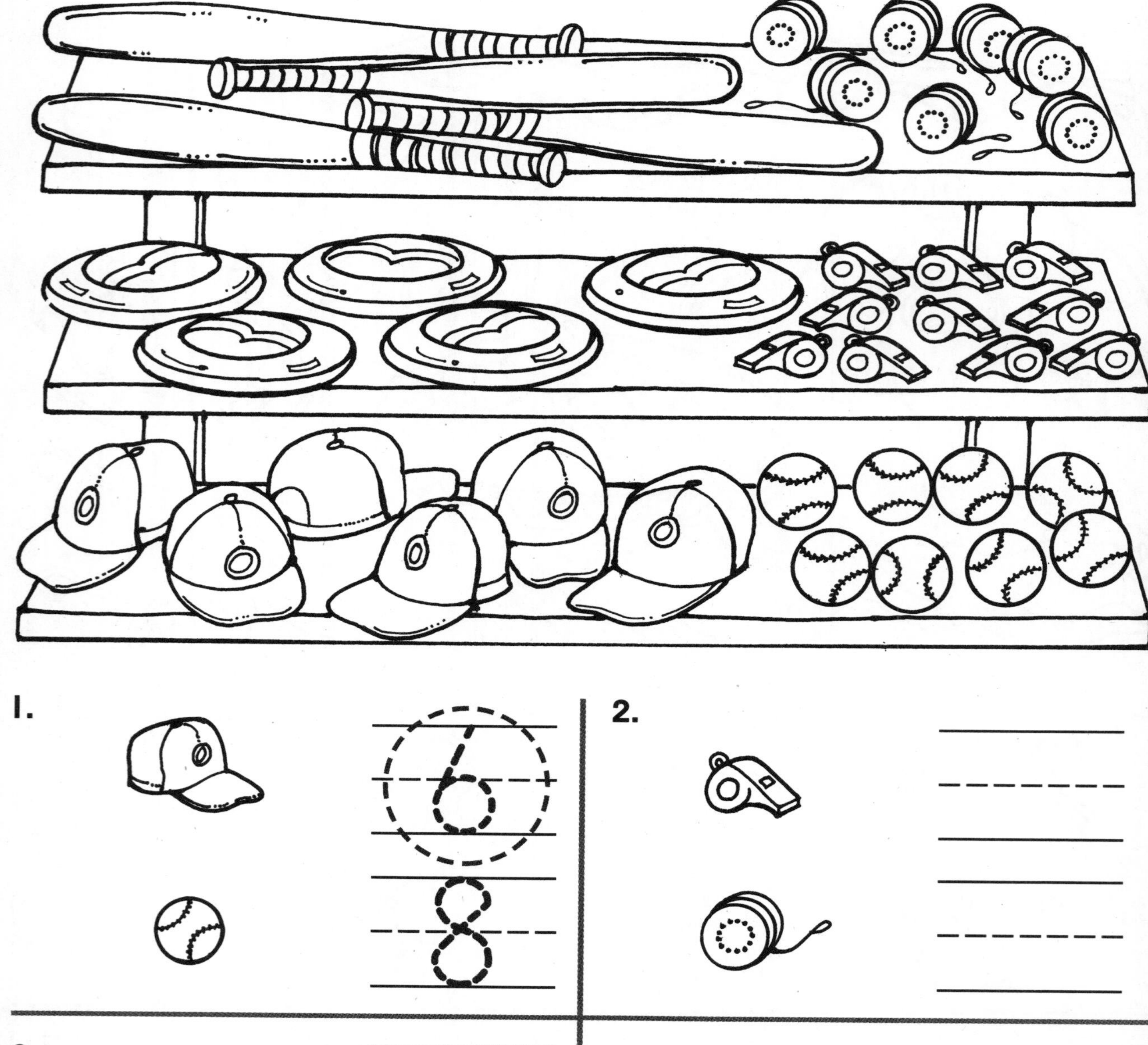

1. 6 8

2.

3.

4.

Count the objects. Write the number.
Circle the number that is less.

Name ______________________

GETTING READY 8

Order Through 10

1.

2.

0	1	2				6			9	

3.

0				4			7			10

1. Connect the dots in order.
2.–3. Write the missing numbers.

Name ________________________________

GETTING READY 9

Ordinal Numbers

second	third	fourth	fifth	first

Cut out the words.
Paste them in order on the rabbits' pails. The winner is first.

Name ______________________________

Modeling Addition Story Problems

Color to make two groups.
Write how many there are in each.
Write how many there are in all.

1.

2 ______ 3 ______ 5 ______ in all

2.

______ ______ ______ in all

3.

______ ______ ______ in all

4.

______ ______ ______ in all

Name ____________________

Adding 1

Draw and color 1 more in each group.

Write the sum.

1. 5 + 1 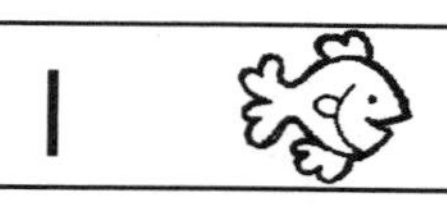 = 6 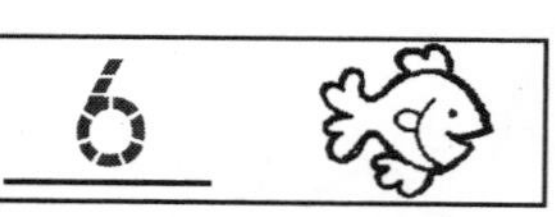

2. 2 + 1 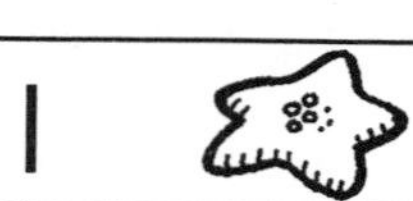 = ____ 

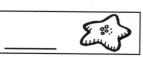

3. 3 + 1 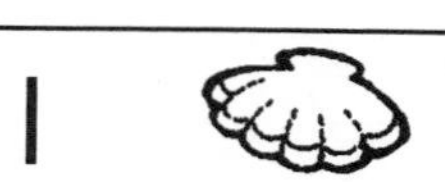 = ____ 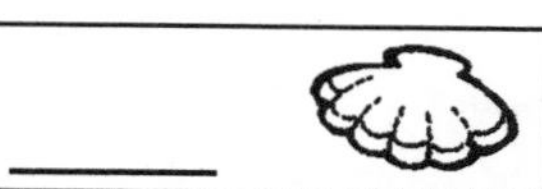

4.  + 1 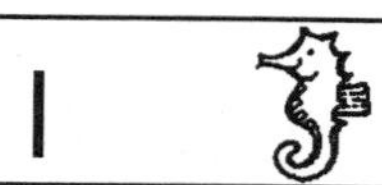 = ____ 

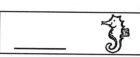

Name ______________________________

LESSON 1.3

Adding 2

Draw and color 2 more in each group.

Write the sum.

1. 4 + 2  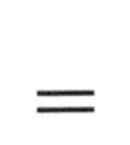 =  6 

2. 3 + 2  = ____ 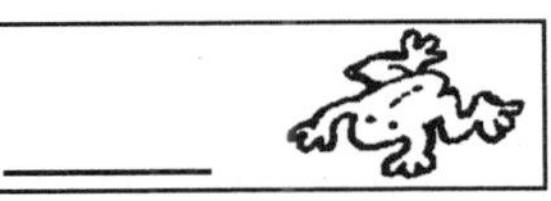

3. 2 + 2 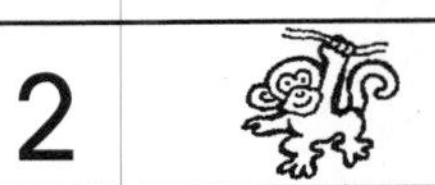 = ____ 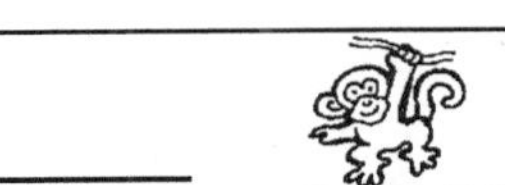

4. 1 + 2  = ____ 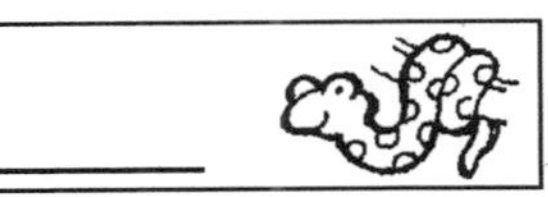

Name ______________________________

LESSON 1.4

Using Pictures to Add

Write the sum.

1.

$1 + 3 = $ 4

2.

$3 + 3 = $ ____

3.

$2 + 1 = $ ____

4.

$1 + 1 = $ ____

5.

$3 + 2 = $ ____

6.

$4 + 2 = $ ____

Name ____________________

Writing Addition Sentences

Write a number in each □.
Write + or = in each ○.
Write the sum.

1.

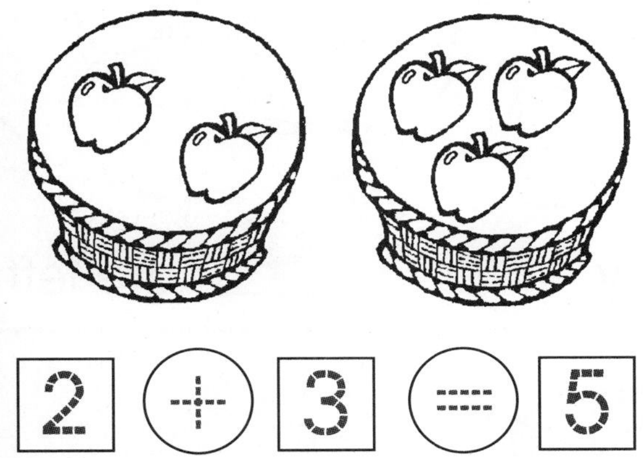

2 + 3 = 5

2.

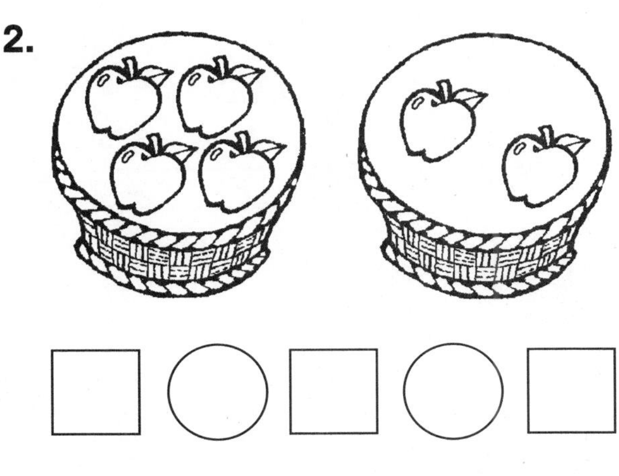

□ ○ □ ○ □

3.

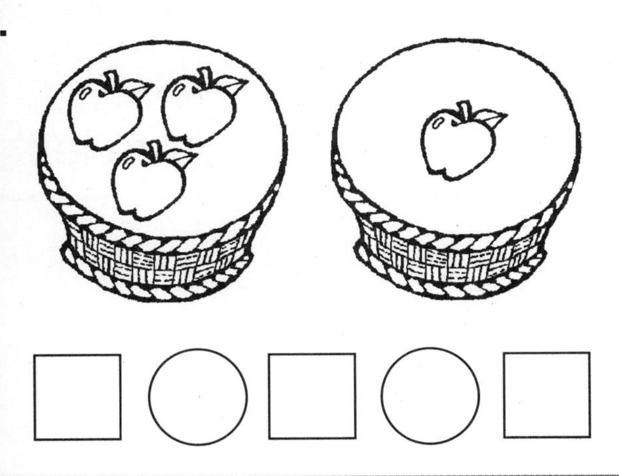

□ ○ □ ○ □

4.

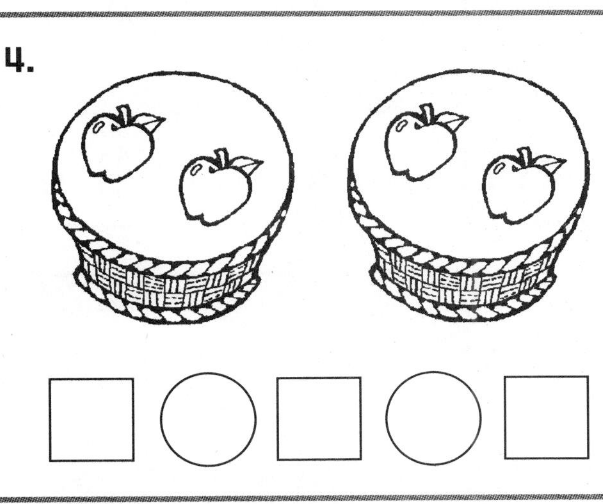

□ ○ □ ○ □

5.

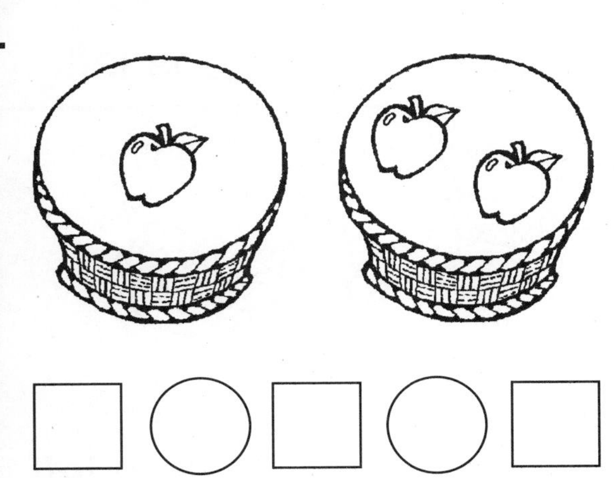

□ ○ □ ○ □

6.

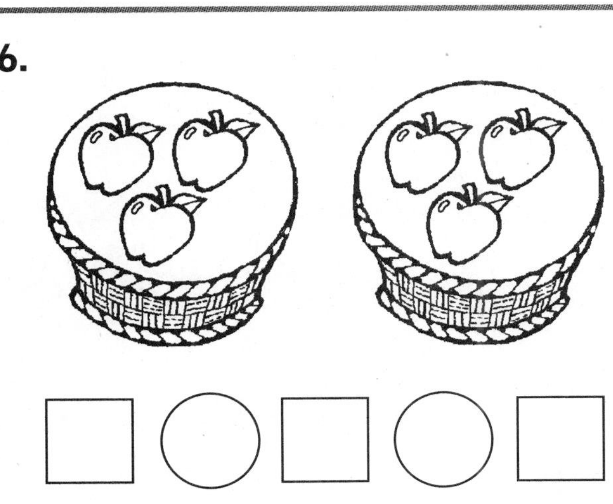

□ ○ □ ○ □

Name ________________________________

LESSON 2.1

Modeling Subtraction Story Problems

Circle the chickens that go away.
Write how many chickens are left.

1.

6 chickens 2 go away __4__ are left

2.

5 chickens 3 go away _____ are left

3.

2 chickens 1 goes away _____ is left

4.

5 chickens 2 go away _____ are left

Name

Subtracting 1

Tell a story to a friend.
Cross out 1 in each picture.
Draw how many are left.

1.

2.

3.

Name ______________________________

Subtracting 2

Draw a picture to show the subtraction sentence.
Write how many are left.

1.

5 − 2 = 3

2.

4 − 2 = ____

3.

3 − 2 = ____

4.

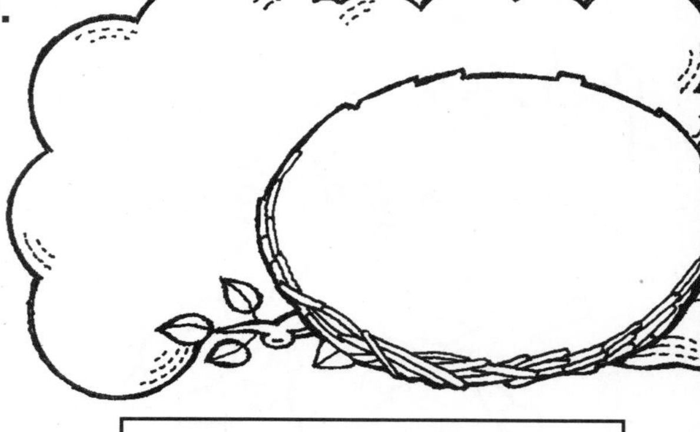

6 − 1 = ____

5.

5 − 1 = ____

6.

2 − 1 = ____

ame ______________________________

Vriting Subtraction Sentences

:ross out some in each row.
Vrite the subtraction sentence.

6 − 1 = 5

____ − ____ = ____

____ − ____ = ____

____ − ____ = ____

____ − ____ = ____

____ − ____ = ____

Name ____________________

Fun in the Sun!

Circle the correct number sentence.
Write the sum or the difference.

1.

$5 + 1 = \underline{6}$ (circled)

$5 - 1 = \_\_\_$

2.

$4 + 2 = \_\_\_$

$4 - 2 = \_\_\_$

3.

$3 - 1 = \_\_\_$

$3 + 1 = \_\_\_$

4.

$6 - 3 = \_\_\_$

$6 + 3 = \_\_\_$

ame ______________________________

last Off!

Vrite the sums. Circle each animal
ıat has two sums that are the same.
raw those animals in your spaceship.

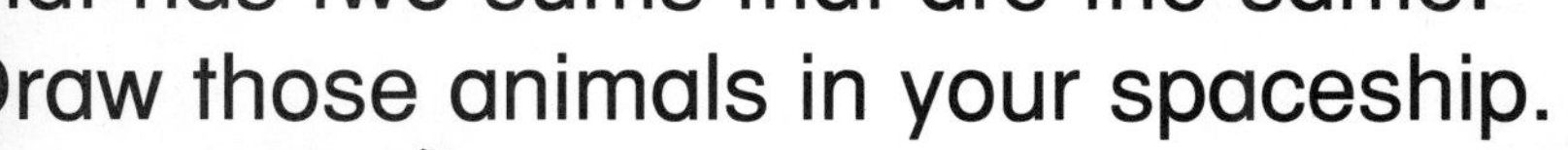

5 + 3 = ____

3 + 5 = ____

2 + 3 = ____

3 + 3 = ____

3 + 4 = ____

4 + 3 = ____

2 + 4 = ____

4 + 2 = ____

2 + 2 = ____

2 + 7 = ____

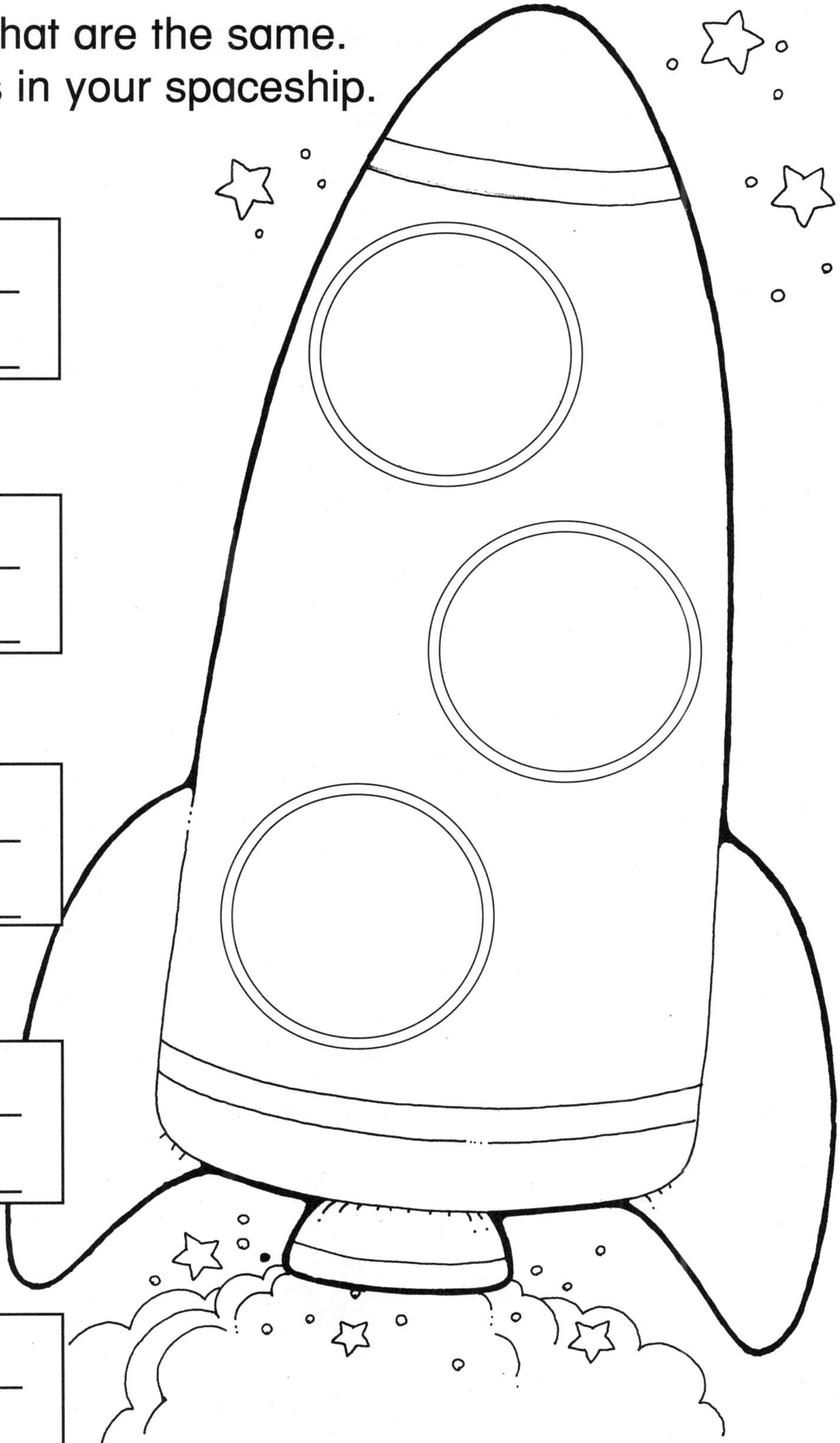

Name ______________________________

Addition Combinations

Draw things to make the sum in two ways.
Complete the addition sentences.

1.

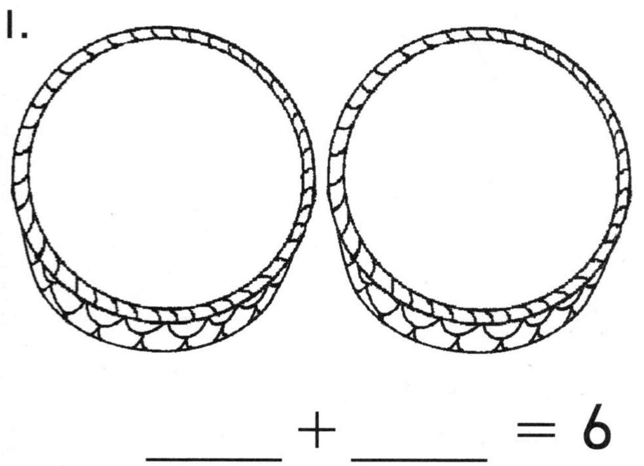

____ + ____ = 6

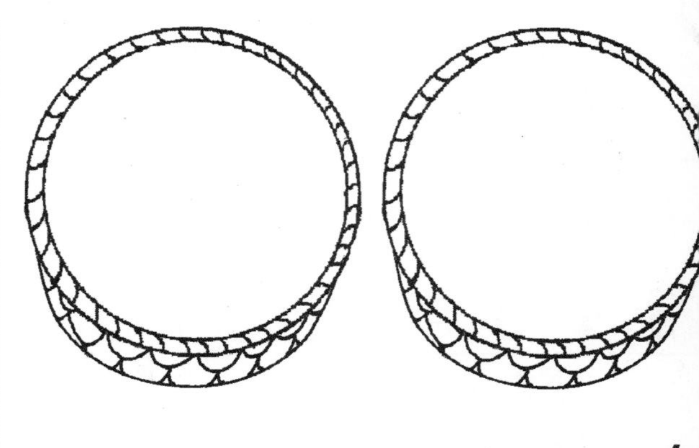

____ + ____ = 6

2.

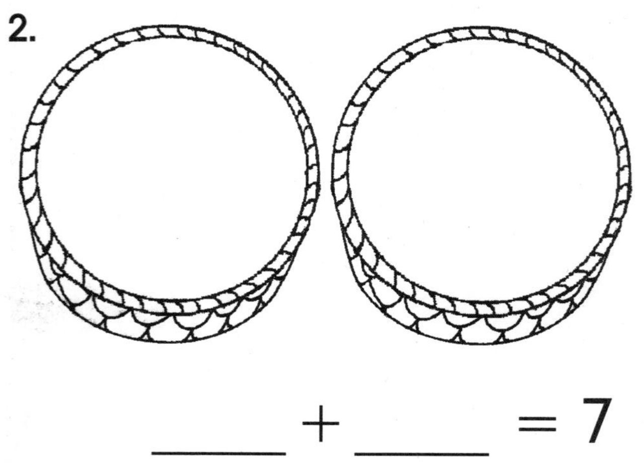

____ + ____ = 7

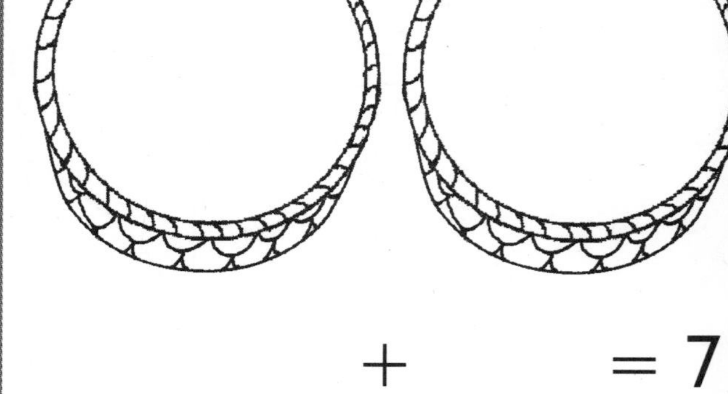

____ + ____ = 7

3.

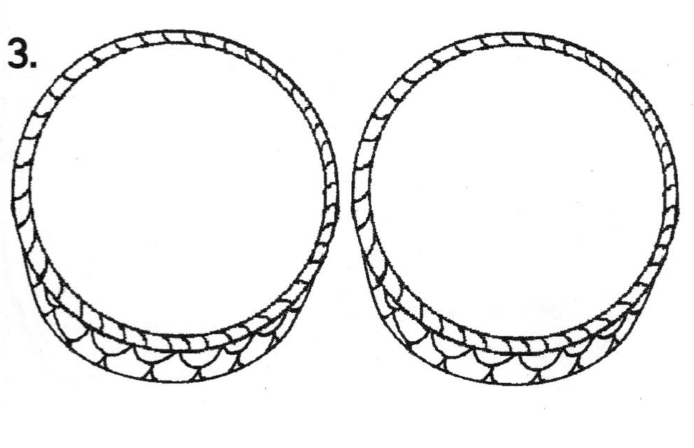

____ + ____ = 8

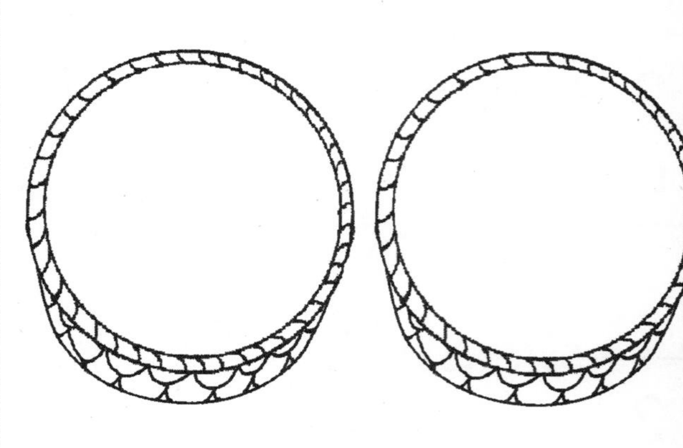

____ + ____ = 8

Name ______________________

More Addition Combinations

Circle the ways to make each sum.

7	8	9	10
6 + 1	2 + 4	9 + 0	9 + 1
5 + 3	4 + 4	2 + 7	7 + 3
2 + 5	6 + 1	3 + 5	2 + 6
3 + 3	7 + 1	8 + 1	6 + 4
2 + 4	4 + 3	4 + 5	2 + 8
7 + 0	3 + 5	1 + 7	5 + 3
0 + 5	6 + 2	2 + 4	3 + 7
7 + 2	0 + 8	4 + 1	5 + 5
4 + 3	2 + 5	6 + 3	0 + 9

Name ____________________

LESSON 3.4

Horizontal and Vertical Addition

Circle the two problems that you think have the same sum. Then write the sums to find out.

1. (3 + 4 = 7) 4 + 1 = 5 (3 + 4 = 7, vertical)

2. 2 + 6 = ____ 2 + 4 = ____ 2 + 6 = ____ (vertical)

3. 4 + 6 = ____ 6 + 4 = ____ 6 + 3 = ____ (vertical)

4. 3 + 5 = ____ 2 + 2 = ____ 3 + 5 = ____ (vertical)

5. 3 + 4 = ____ 7 + 2 = ____ 2 + 7 = ____ (vertical)

6. 5 + 0 = ____ 0 + 4 = ____ 0 + 5 = ____ (vertical)

ame ____________________

hopping Trip

Vrite the addition sentence.

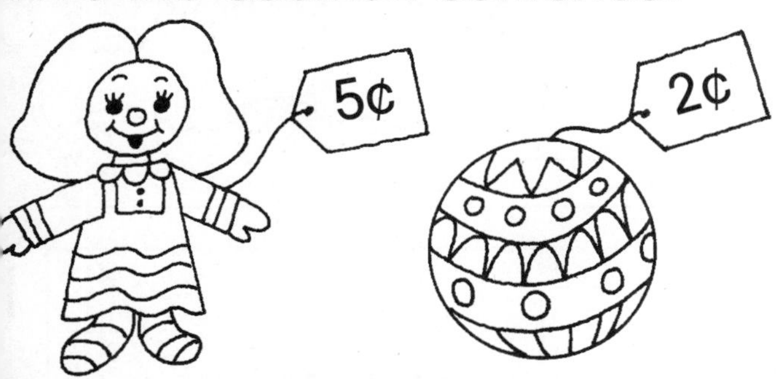

5 ¢ + 2 ¢ = 7 ¢

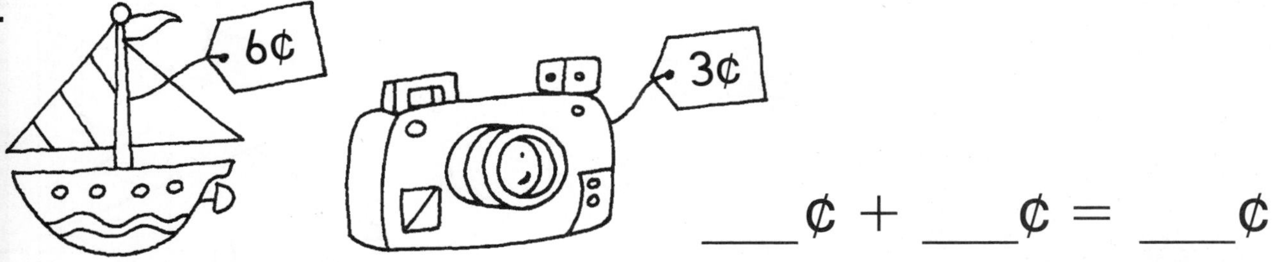

____ ¢ + ____ ¢ = ____ ¢

____ ¢ + ____ ¢ = ____ ¢

____ ¢ + ____ ¢ = ____ ¢

____ ¢ + ____ ¢ = ____ ¢

Name ______________________________

LESSON 4.1

Counting with Colors

Count on to add.
Color blue the spaces
that show the sums.

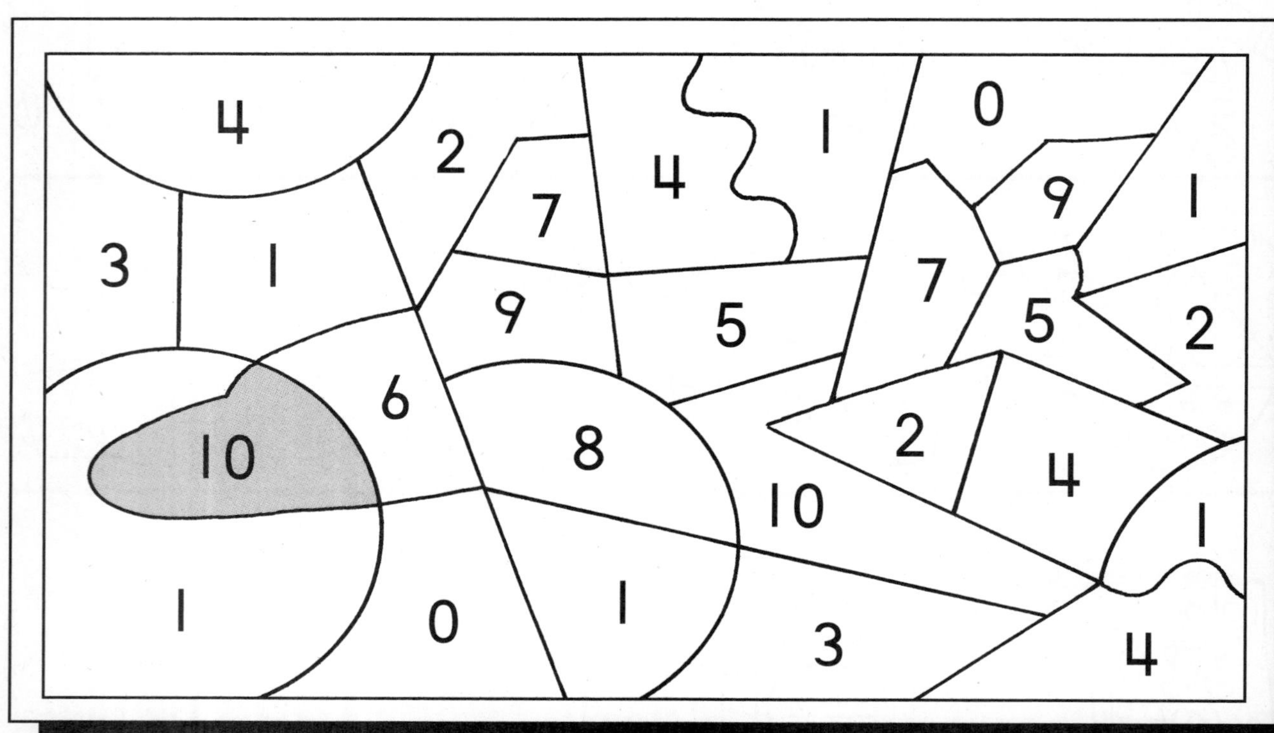

1. $\begin{array}{r} 9 \\ +\ 1 \\ \hline 10 \end{array}$ $\begin{array}{r} 7 \\ +\ 2 \\ \hline \end{array}$ $\begin{array}{r} 6 \\ +\ 1 \\ \hline \end{array}$ $\begin{array}{r} 4 \\ +\ 1 \\ \hline \end{array}$ $\begin{array}{r} 6 \\ +\ 2 \\ \hline \end{array}$

2. $\begin{array}{r} 5 \\ +\ 2 \\ \hline \end{array}$ $\begin{array}{r} 4 \\ +\ 1 \\ \hline \end{array}$ $\begin{array}{r} 7 \\ +\ 2 \\ \hline \end{array}$ $\begin{array}{r} 5 \\ +\ 1 \\ \hline \end{array}$ $\begin{array}{r} 8 \\ +\ 2 \\ \hline \end{array}$

Name ____________________

LESSON 4.2

Counting with Flower Power

Count on to find each sum.

1.

2.

3.

4.

Name ______________________________

Double Bubbles

Add. Color the doubles red.

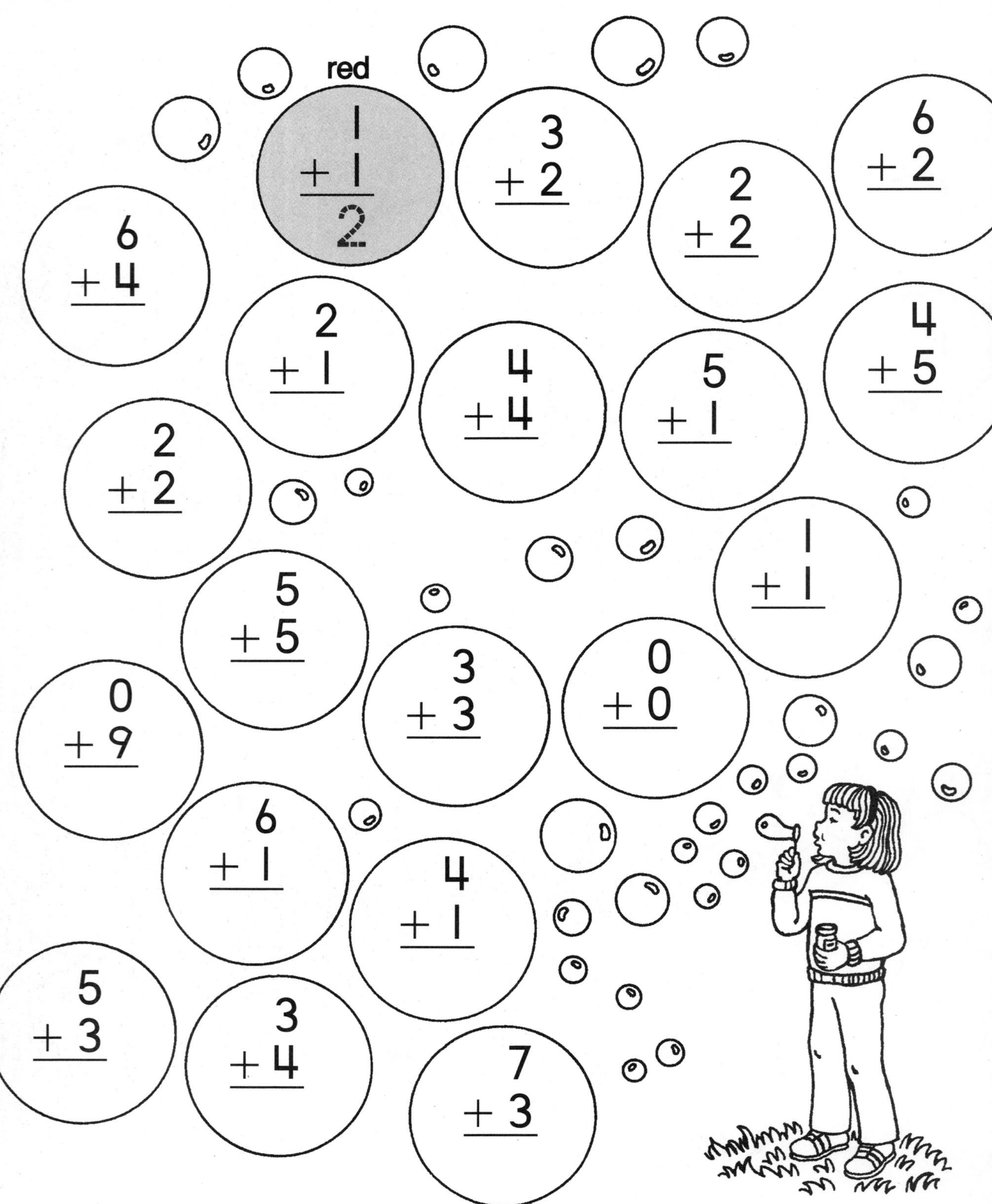

Name ______________________________

LESSON 4.4

Adding Around the House

Write the sum.

1.

2.

3.

4.

Name ______________________

Problem Solving • Write a Number Sentence

Solve. Write the number sentence.

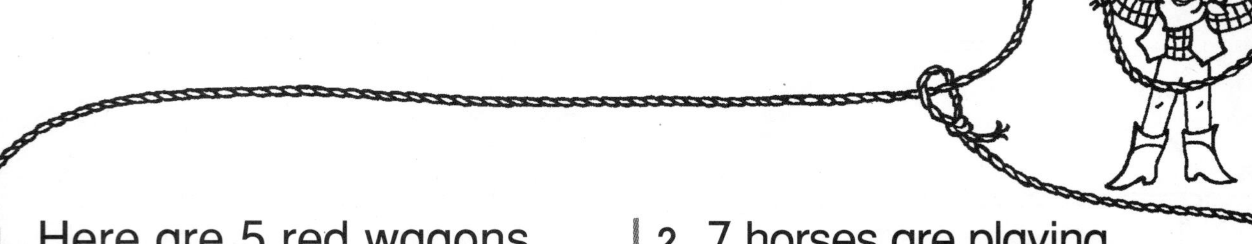

1. Here are 5 red wagons.
Here is 1 blue wagon.
How many wagons in all?

5 (+) 1 = 6

2. 7 horses are playing.
3 run away.
How many are left?

____ ◯ ____ = ____

3. 6 cows eat grass.
1 goes away.
How many cows are left?

____ ◯ ____ = ____

4. Lennie has 3 hats.
Diana has 4 hats.
How many hats in all?

____ ◯ ____ = ____

5. 2 boots are black.
6 boots are brown.
How many boots in all?

____ ◯ ____ = ____

6. There are 8 girls.
5 walk away.
How many girls are left?

____ ◯ ____ = ____

Name ______________________________

LESSON 5.1

Tree Time!

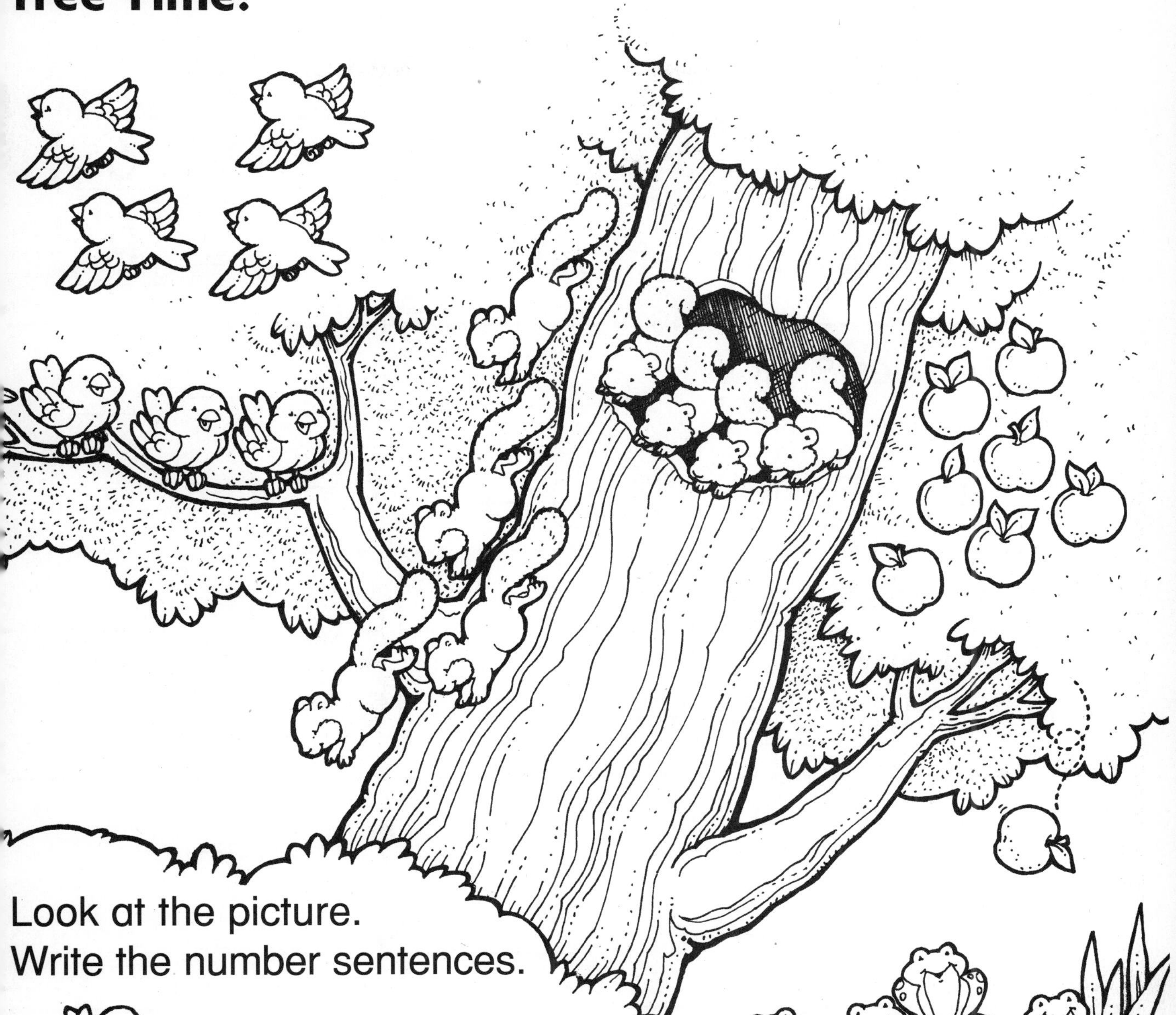

Look at the picture.
Write the number sentences.

1.  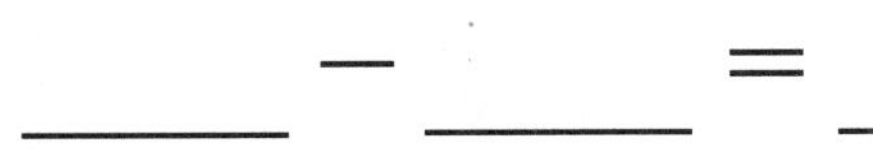

2.  

3.  ______ − ______ = ______

4.  ______ − ______ = ______

Name ______________________________

LESSON 5.2

More Subtraction Combinations

Solve the problem. Circle the correct number sentence.

1. Ruben has 10 .
He eats 4.
How many  are left? 6

(circled) 10 − 4 = 6
10 − 2 = 8

2. Martha sees 9 .
2 hop away.
How many  are left? ____

9 − 4 = ____
9 − 2 = ____

3. Jeffrey sails 10 .
2 float away.
How many are left? ____

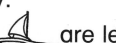

10 − 2 = ____
10 − 3 = ____

4. Beth has 9 .
She spends 4.
How many  are left? ____

9 − 4 = ____
9 − 7 = ____

5. Ray sees 9 .
He picks 5.
How many  are left? ____

9 − 5 = ____
9 − 3 = ____

6. Suki has 10 .
She gives 1 away.
How many  are left? ____

10 − 6 = ____
10 − 1 = ____

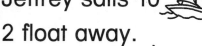

Name ______________________

Find the Flags

Write the sum. Color the flags.

Color 5 blue. Color 7 yellow.

Color 6 red. Color 1 green.

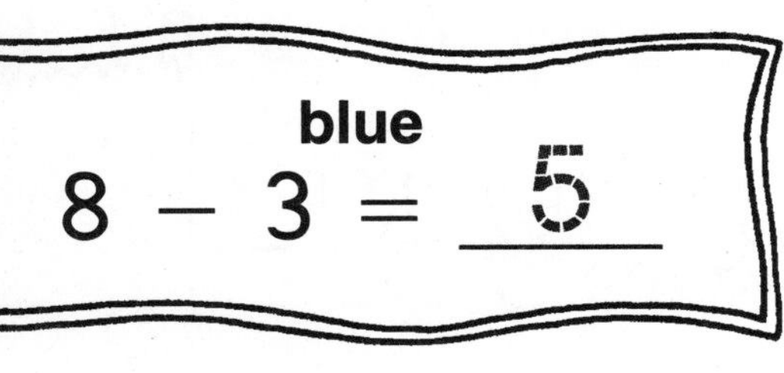

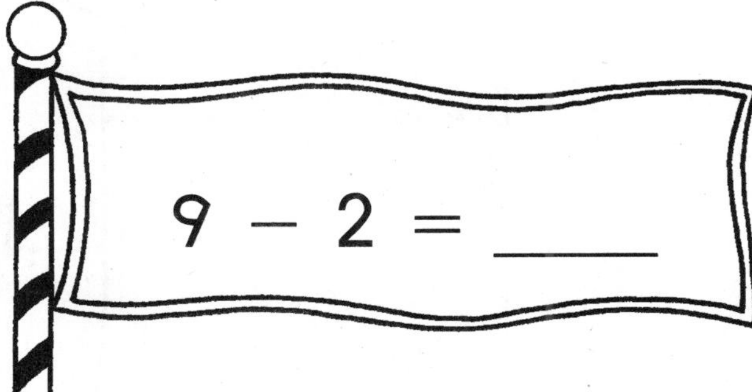

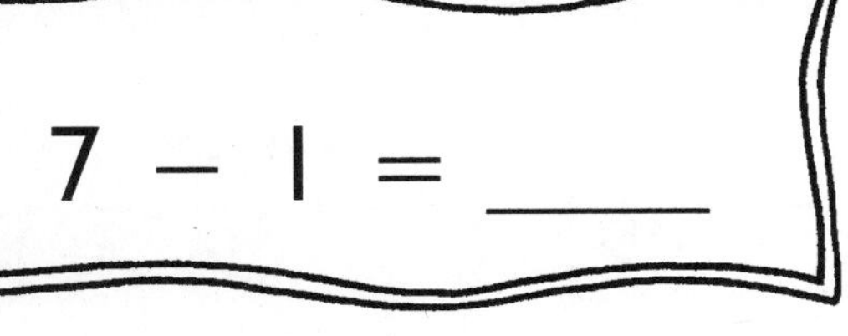

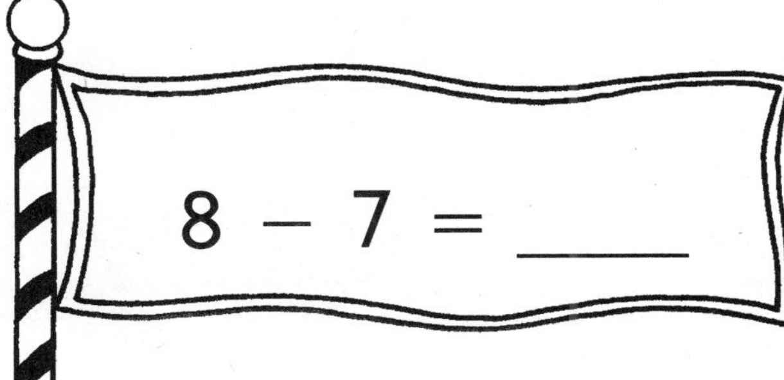

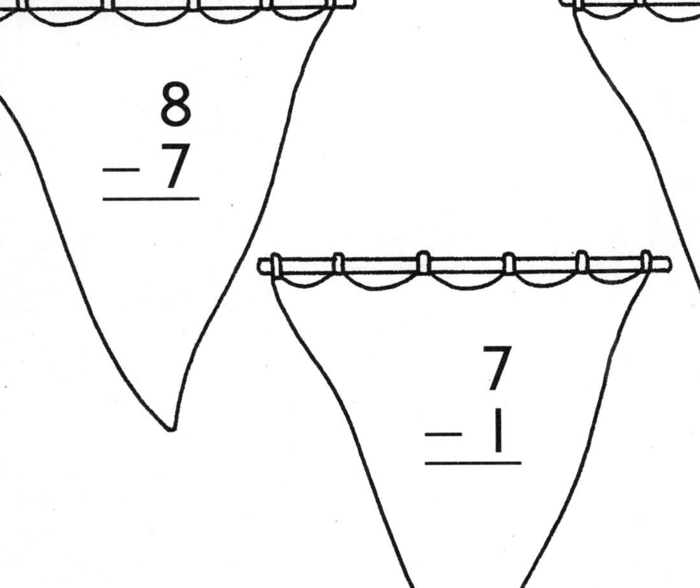

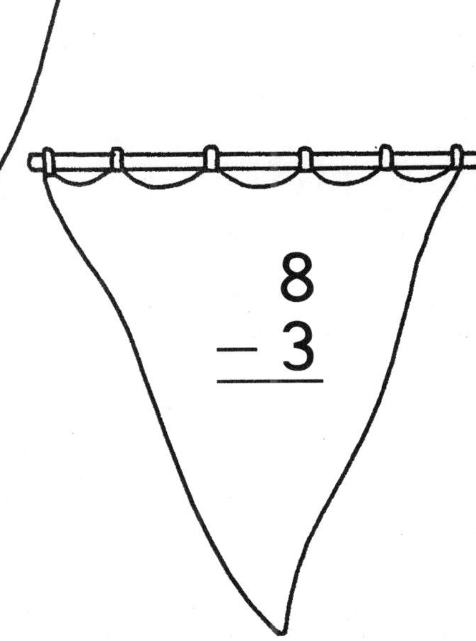

Name ______________________________

Follow the Path

Follow each path.
Complete the facts in each fact family.

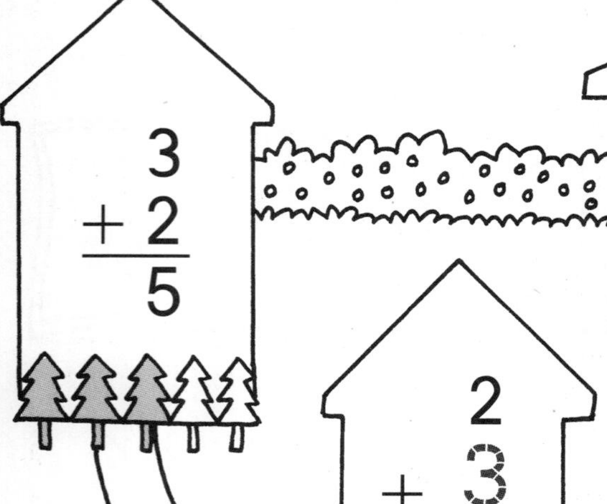

1 + 6 = 7

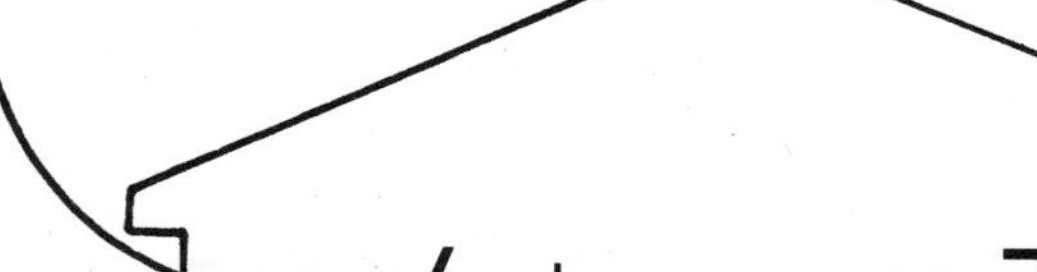

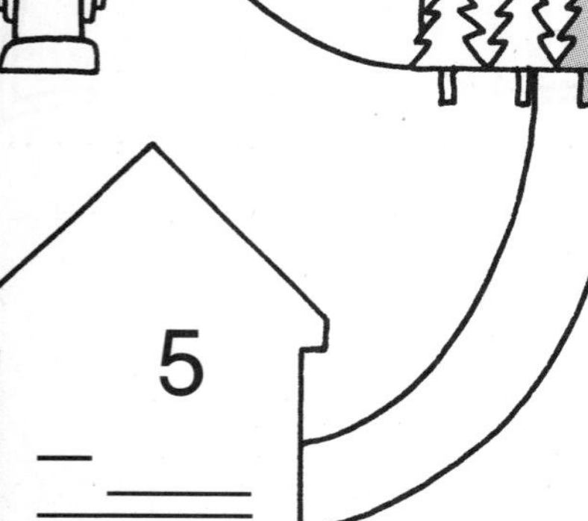

7 − ___ = ___

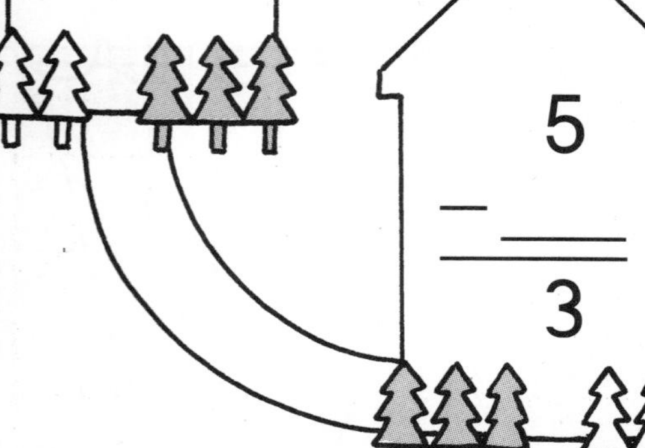

___ − ___ = ___

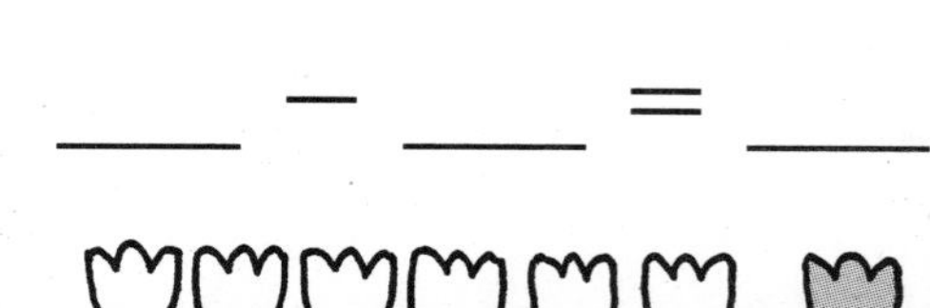

Name ______________________

Subtracting to Compare

Solve the problem. Write how many more.

1. There are 7 .

   There are 5 .

   7 – 5 = 2

   2 more 

2. There are 9 .

   There are 2 .

   ____ – ____ = ____

   ____ more 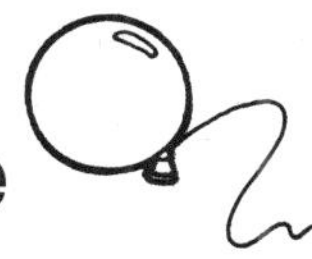

3. There are 10 .

   There are 4 .

   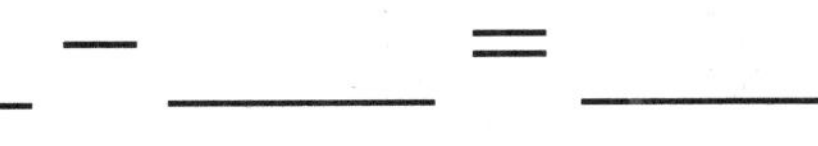

   ____ – ____ = ____

   ____ more 

4. There are 8 .

   There are 6 .

   ____ – ____ = ____

   ____ more 

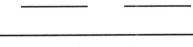

Name ______________________________

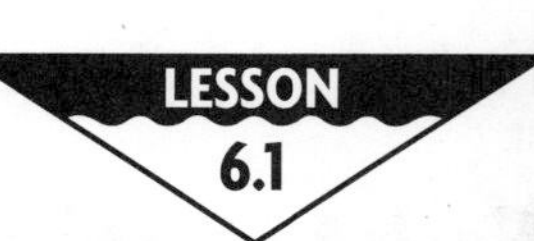

Counting Back 1 and 2

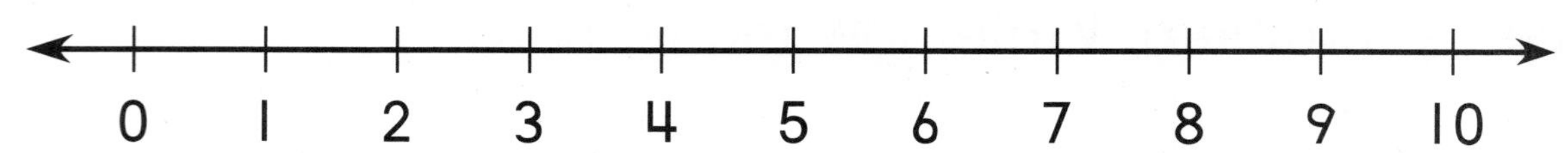

Pick a number.	Subtract.	Write the difference.
1. 9	− 1	= 8
2. ____	− 1	= ____
3. ____	− 2	= ____
4. ____	− 2	= ____
5. ____	− 1	= ____
6. ____	− 2	= ____
7. ____	− 2	= ____
8. ____	− 1	= ____

Name ________________________________

Where Am I?

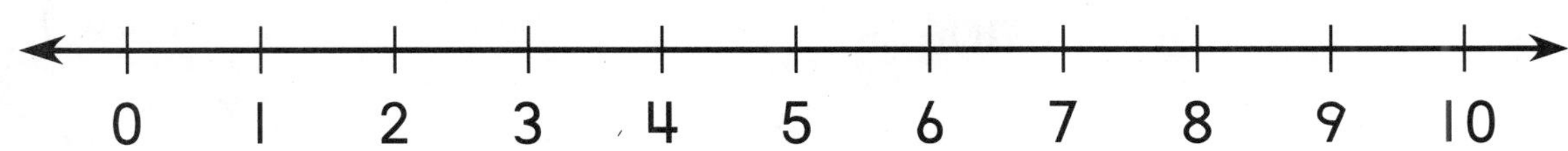

Use the number line. Solve.

1. I am at 4.
I jump back 3.
Where am I?
______

2. I am at 7.
I jump forward 3.
Where am I?
______

3. I am at 3.
I jump forward 1.
Where am I?
______

4. I am at 5.
I jump back 2.
Where am I?
______

5. I am at 8.
I jump forward 2.
Where am I?
______

6. I am at 6.
I jump back 1.
Where am I?
______

7. I am at 10.
I jump back 3.
Where am I?
______

8. I am at 2.
I jump back 2.
Where am I?
______

9. I am at 9.
I jump back 3.
Where am I?
______

10. I am at 5.
I jump forward 3.
Where am I?
______

Name ______________________________

Fact Flowers

Subtract. Fill in the numbers.

Name ______________________________

Facts Practice

Write the sum or the difference.

Color each 5 red.

Color each 7 blue.

Color each 4 yellow.

Color each 3 green.

Name ______________________________

LESSON 6.5

Problem Solving • Make a Model

Use counters to model the problem. Solve.

1. There are 10 markers.
   Jon takes 6.
   Lindsay takes 1.
   How many markers are left?

   3 markers

2. There are 9 paintbrushes.
   Jamie takes 1.
   Paul takes 1.
   How many are left?

   ______ paintbrushes

3. Ned has 3 crayons.
   Bill has 2 crayons.
   Corey has 1 crayon.
   How many crayons in all?

   ______ crayons

4. Ann has 8 pencils.
   She gives 2 to Mary.
   She gives 2 to Sean.
   How many are left?

   ______ pencils

5. Michele has 5 stickers.
   Heather has 1 sticker.
   Jordan has 2 stickers.
   How many stickers in all?

   ______ stickers

6. Jessica has 3 books.
   Lisa has 1 book.
   Sam has 1 book.
   How many books in all?

   ______ books

Name ______________________________

Solid Figures

Color the rectangular prisms brown.

Color the spheres yellow.

Color the cones blue.

Name ____________________

More Solid Figures

Mark an **X** on the figures that do not belong.

1.

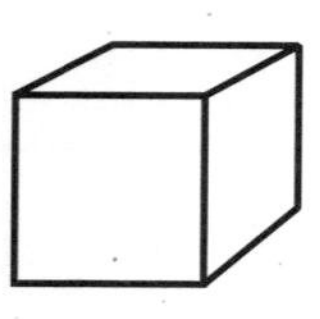
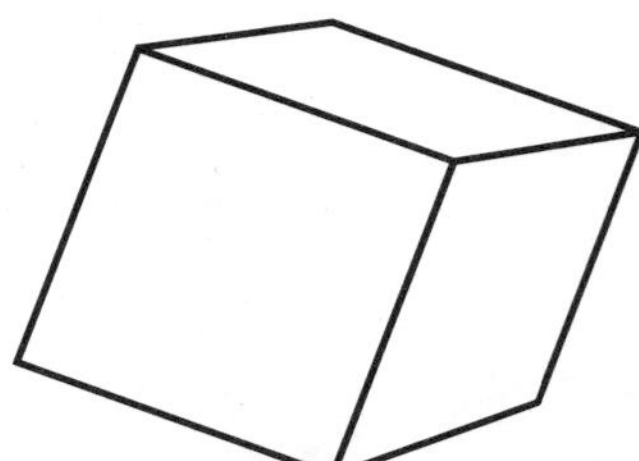
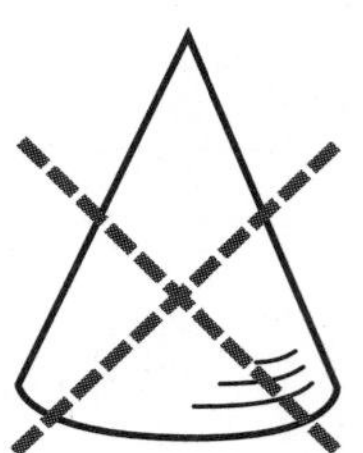
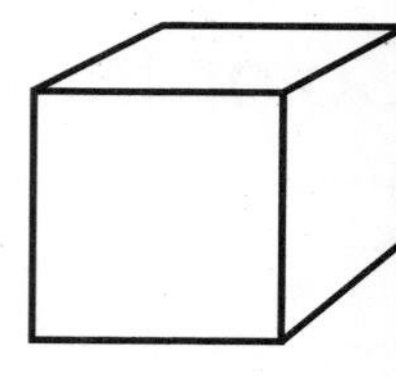

2.

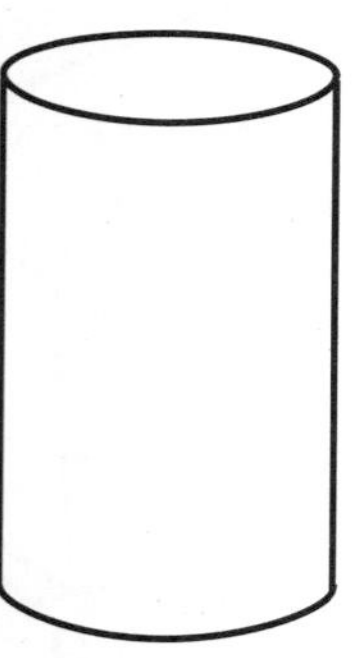
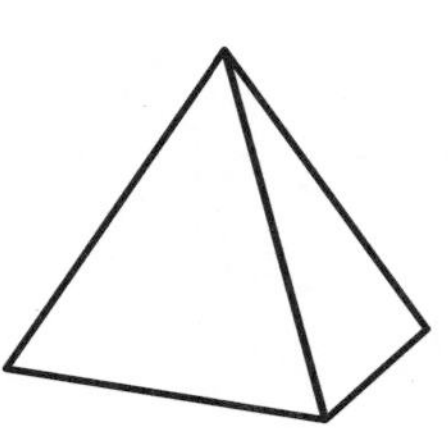

3.

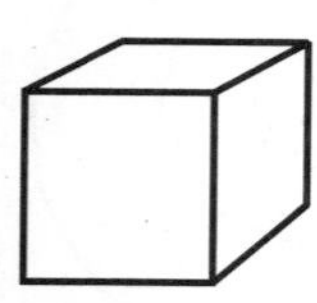
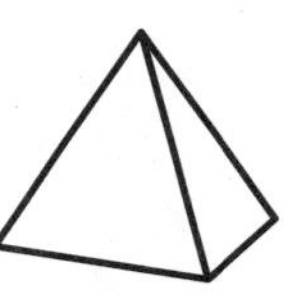
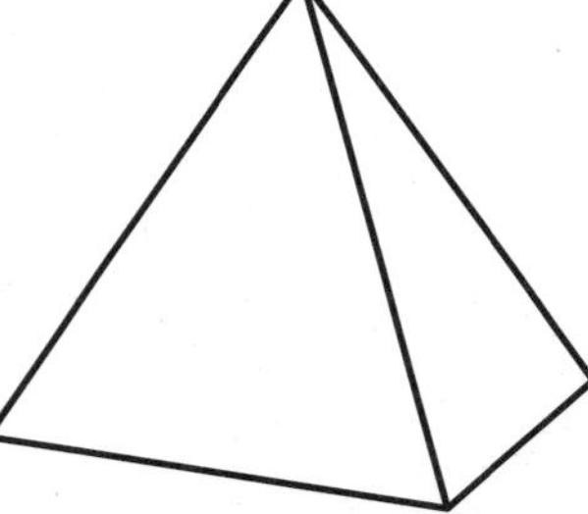
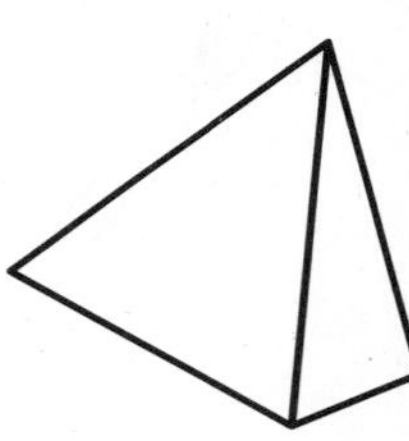

4.

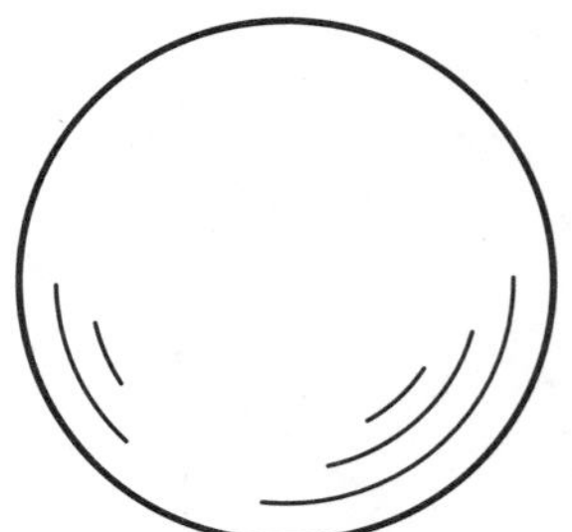

Name ____________________

Sorting Solid Figures

Guess how the figures were sorted.
Write a letter on the prize ribbon.
Use each letter only one time.

A. They can roll.

B. They can slide but not roll.

C. They can stack and slide.

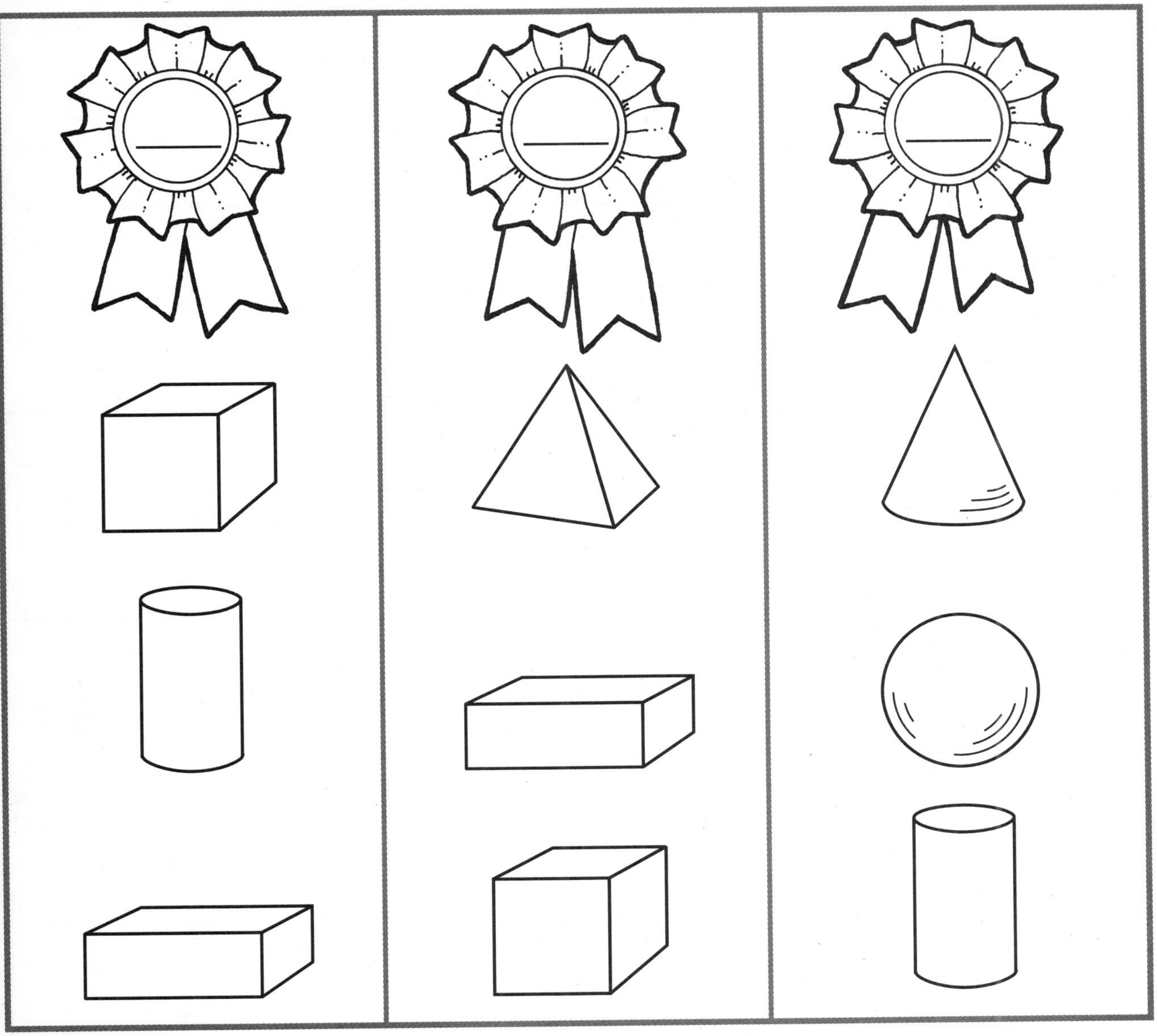

Name ______________________________

How Many Faces?

Circle the answers. Use solid figures to check your answers.

Figure	Are All Faces Flat?		How Many Flat Faces?		
	Yes	No	0	1	4
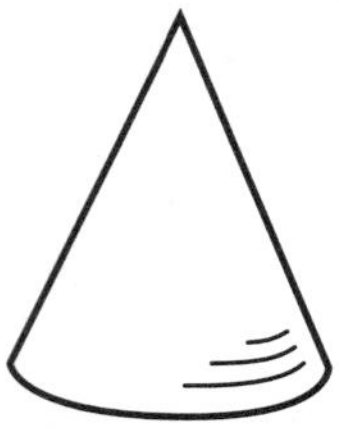	Yes	No	1	2	6
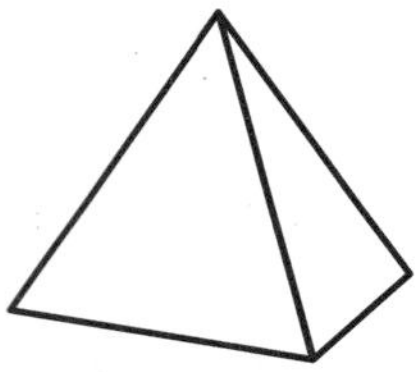	Yes	No	4	5	6
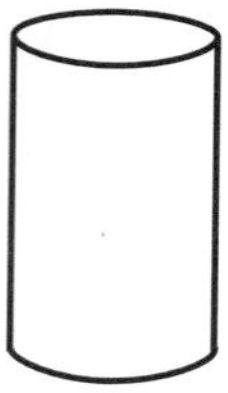	Yes	No	1	2	4
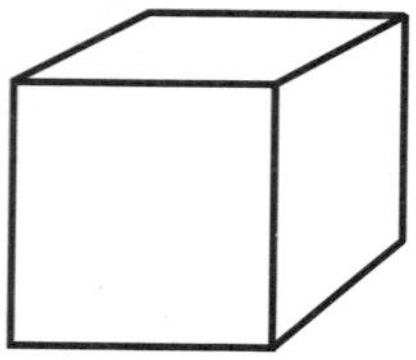	Yes	No	0	4	6
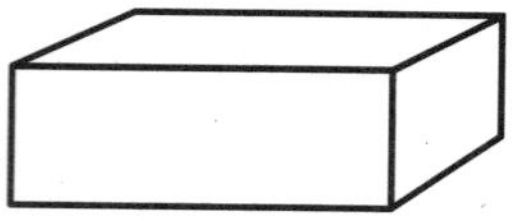	Yes	No	4	6	8

Name ____________________

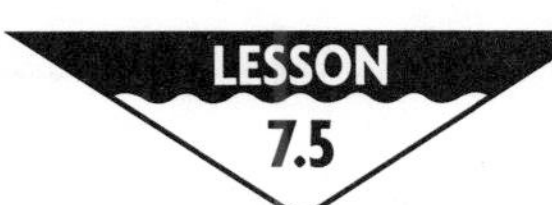

Problem Solving • Make a Model

Circle the figure that is missing from Model B.

Model A | Model B

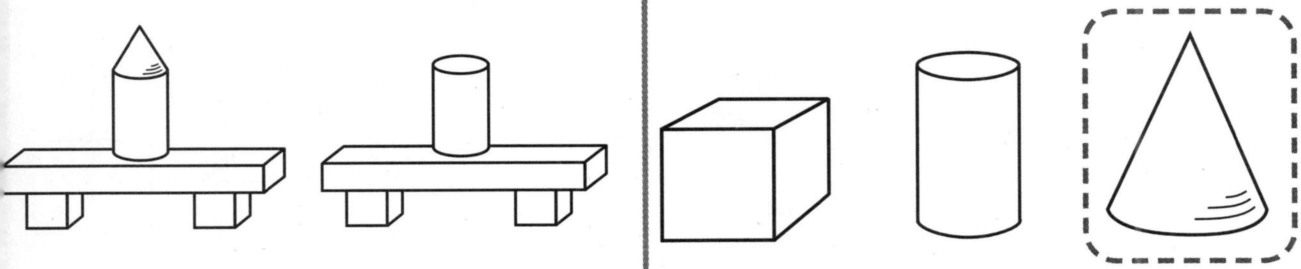

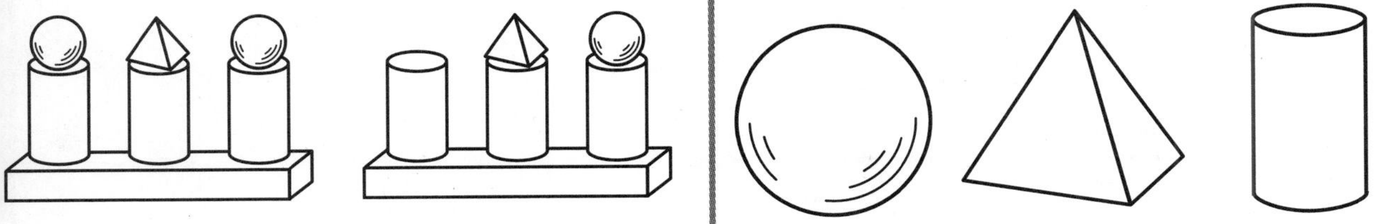

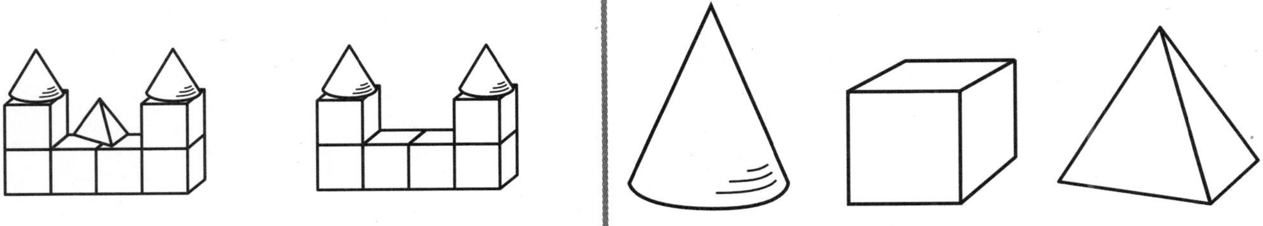

Name ____________________

Plane Figures

Color each □ red.	Color each ○ blue.
Color each △ green.	Color each ▭ yellow.

1.

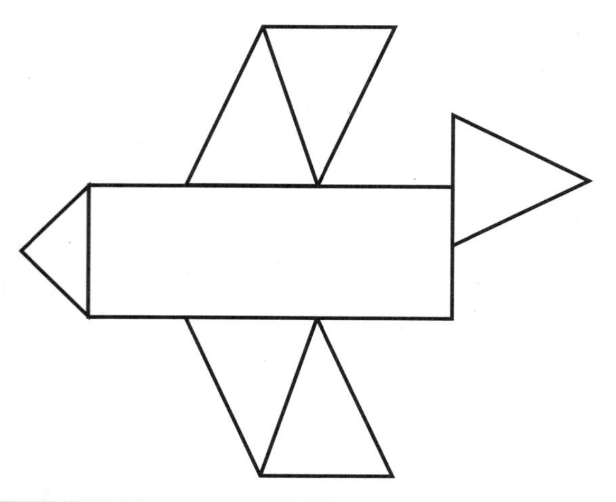

2.

3.

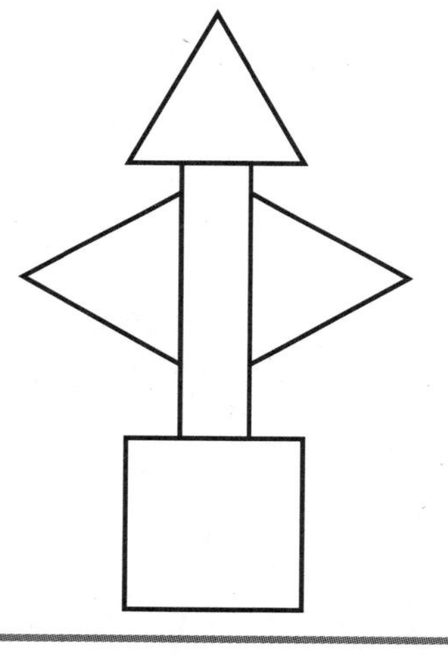

4.

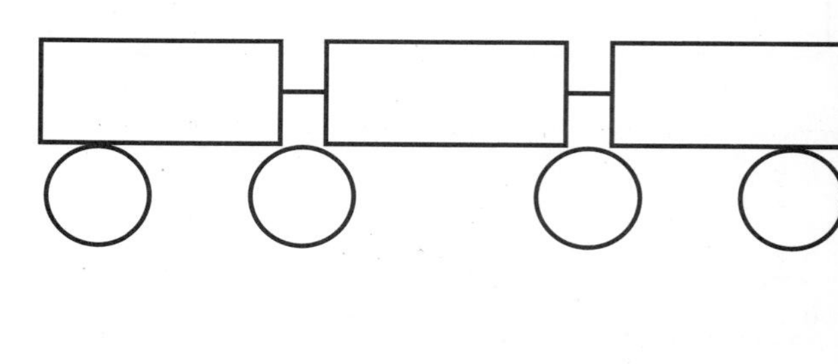

5.

6.

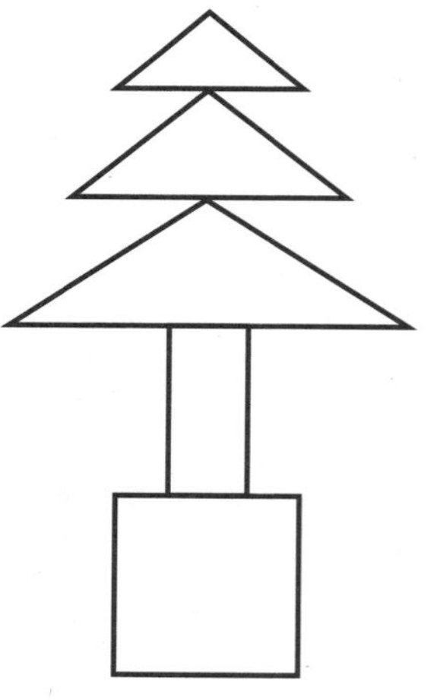

ame ______________________

ides and Corners

ead the riddle.
raw the answer.

. I am a closed figure.
I have 8 sides.
I have 8 corners.
I am a sign that tells cars to stop.

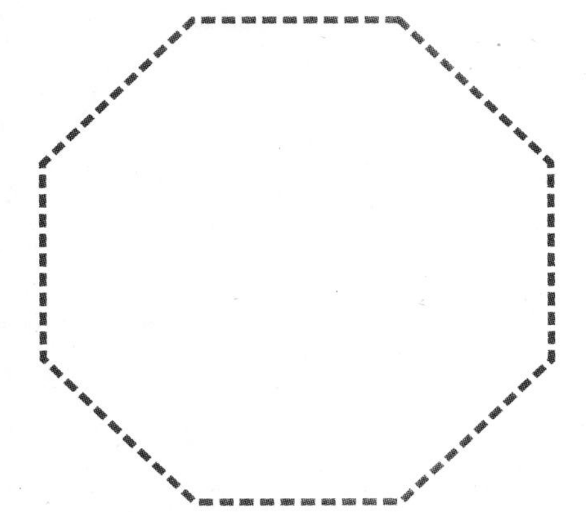

2. I am a closed figure.
I have 10 sides.
I have 10 corners.
You see me in the sky at night.

. I am a closed figure.
I have 7 sides.
I have 7 corners.
I show you which way to go.

4. I am a closed figure.
I have 7 sides.
I have 7 corners.
A king wears me on his head.

Name ____________________

Congruence

Color the figures that are the same size and shape.

1.

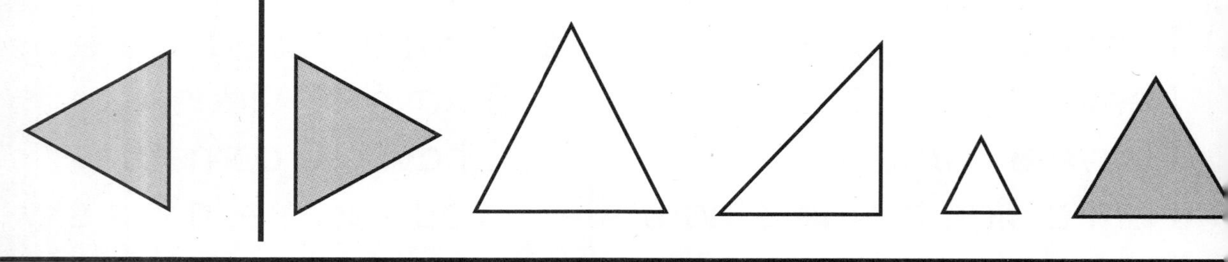

2.

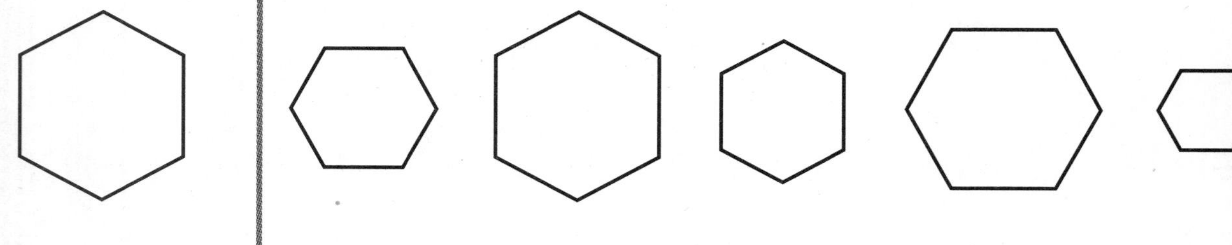

3.

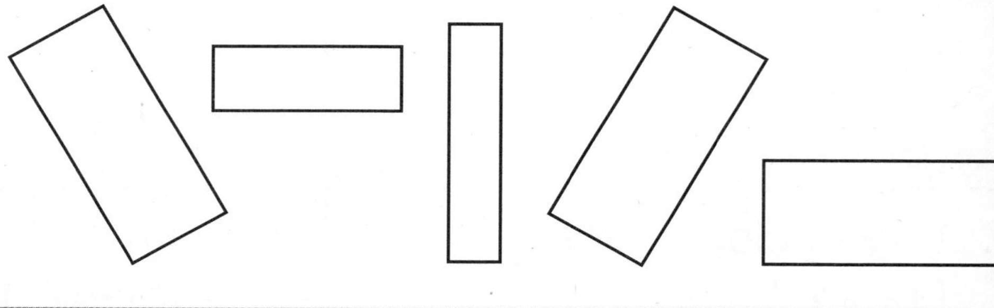

4.

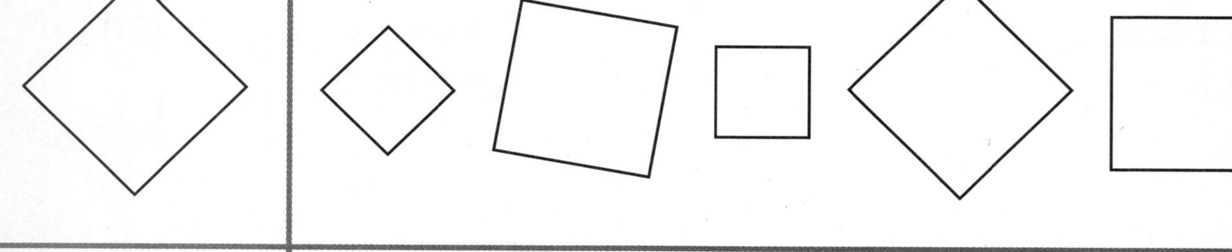

5.

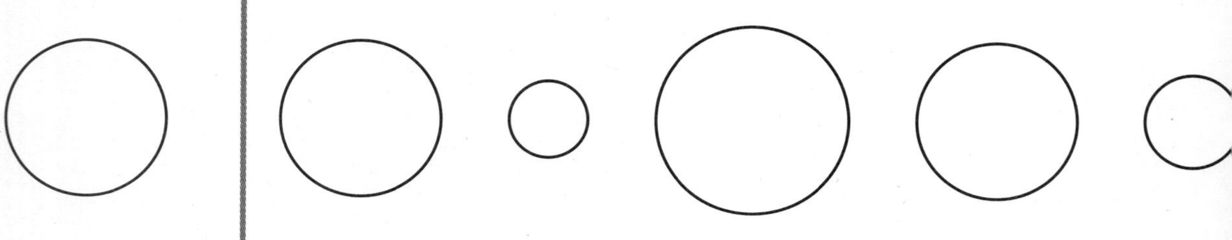

Name ____________________

Symmetry

Draw. Make two sides that match.

1\. 

2\. 

3\. 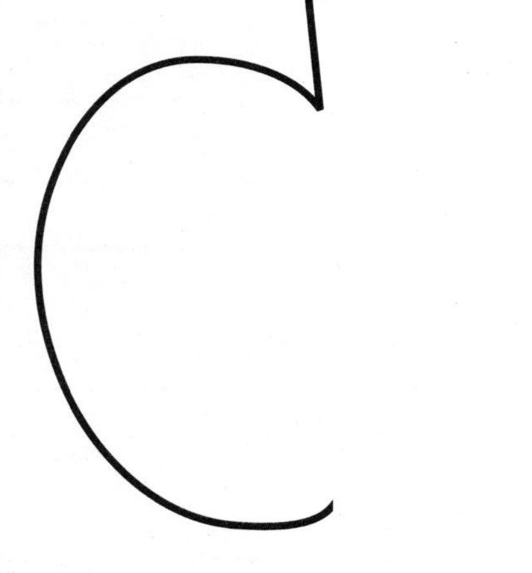

4\. 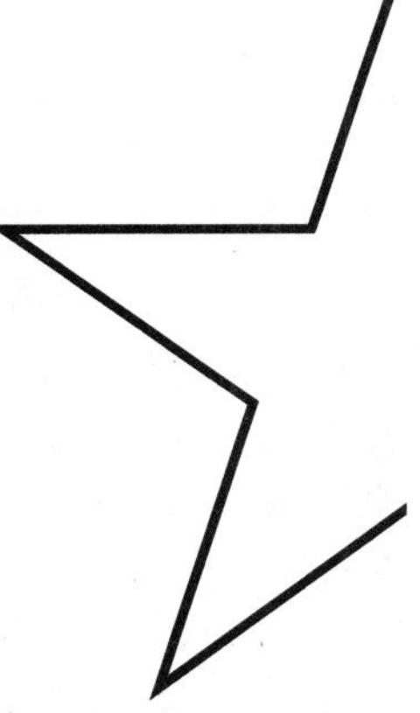

5\. 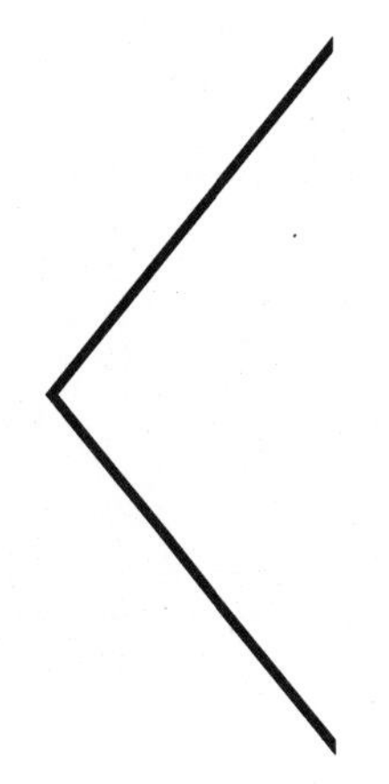

6\. 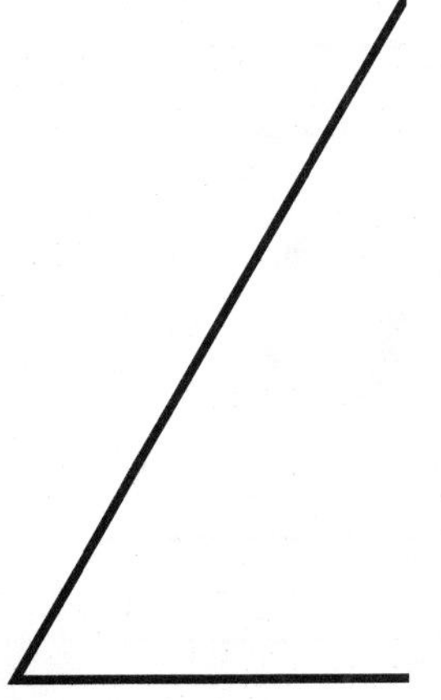

Name ______________________________

Open and Closed

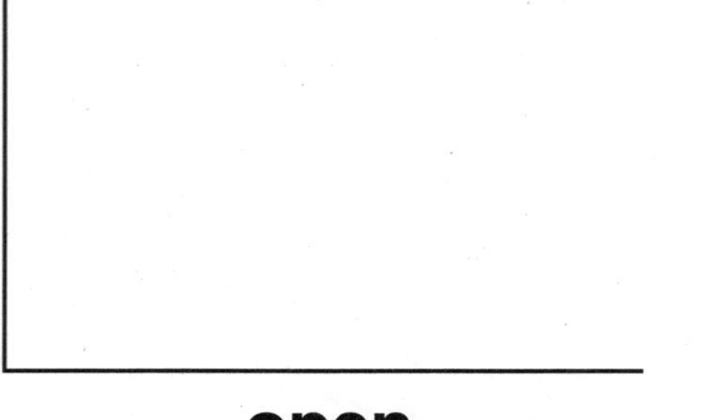

open

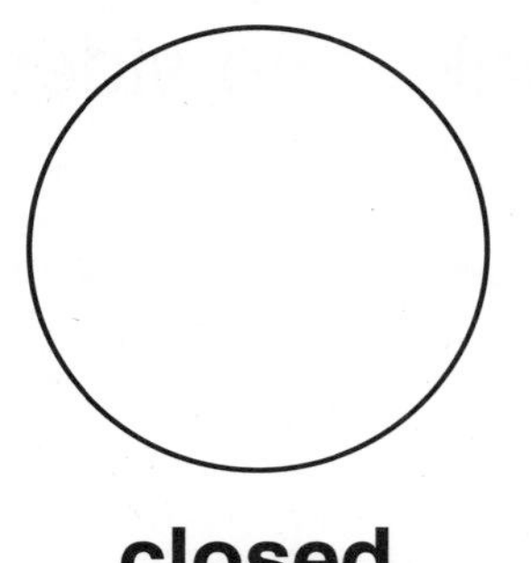

closed

Write **open** or **closed**.

1.

open

2.

3.

4.

5.

6.

Q

Name ___

Inside, Outside, On

Find the correct △ and color it.

1. It is inside circle A and outside circle B.

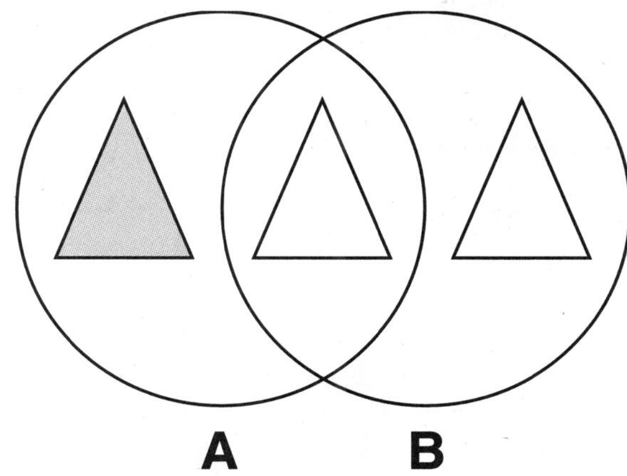

2. It is on circle A and inside circle B.

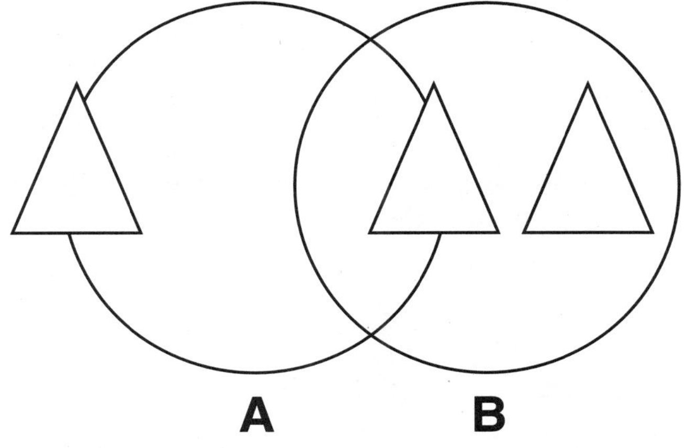

3. It is inside circle A and inside circle B.

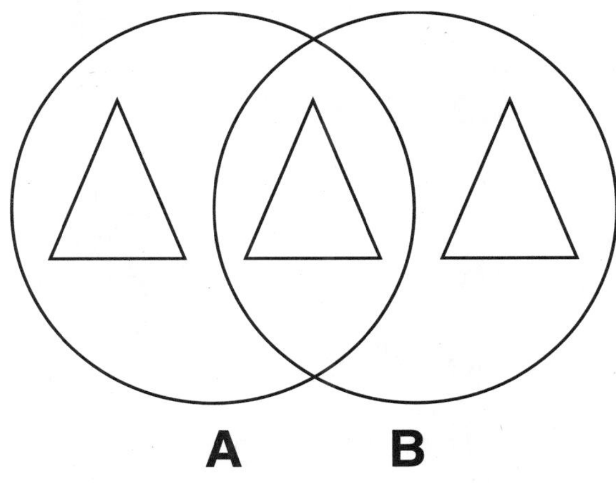

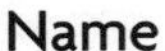

Name ______________________________

Problem Solving • Use a Picture

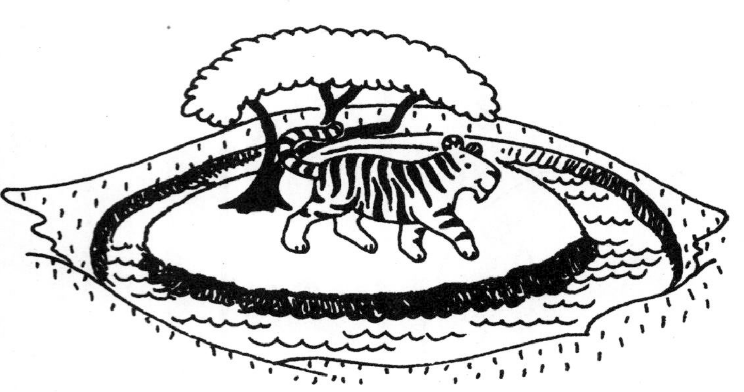

Draw a line to show a path to the gift shop.

1. Start to the right of the bear.
2. Turn left.
3. Turn right at the zebra.
4. Turn right and go to the gift shop.
5. Draw a friend to the left of the bear.

ame

ositions on a Grid

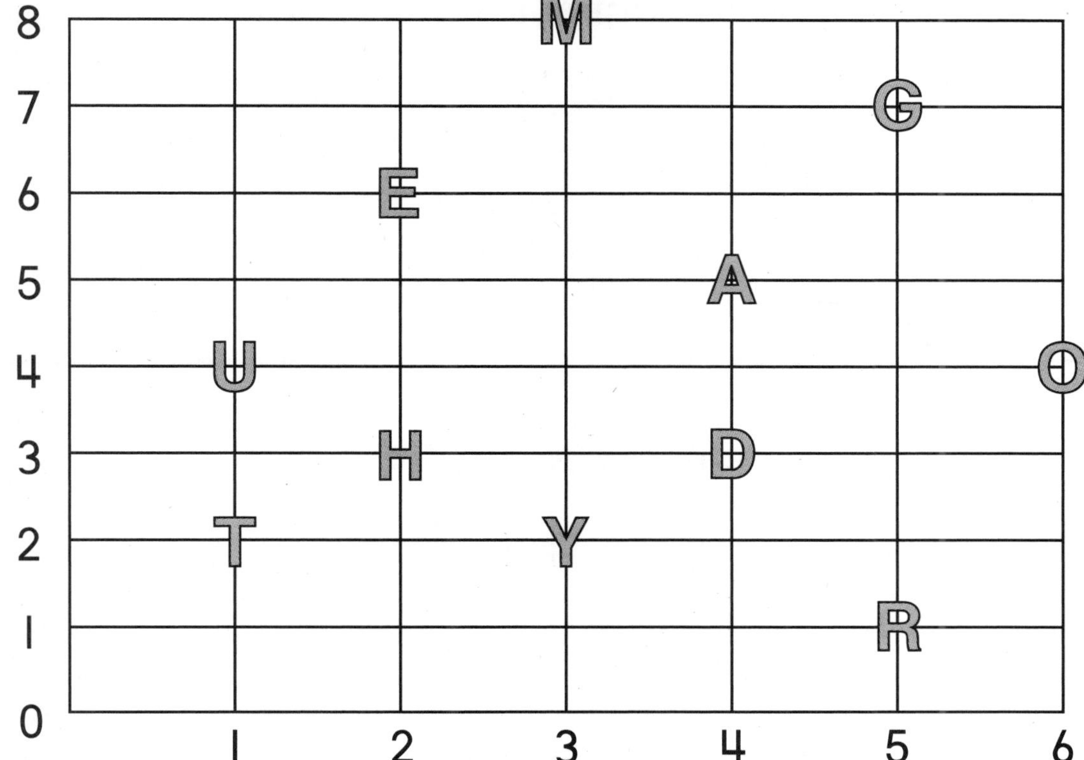

Up

ight

→

Go to the right and then up.
Write the letter. What do the words say?

Name ____________________

LESSON 10.1

Identifying Patterns

Draw the shape to continue the pattern.

1.

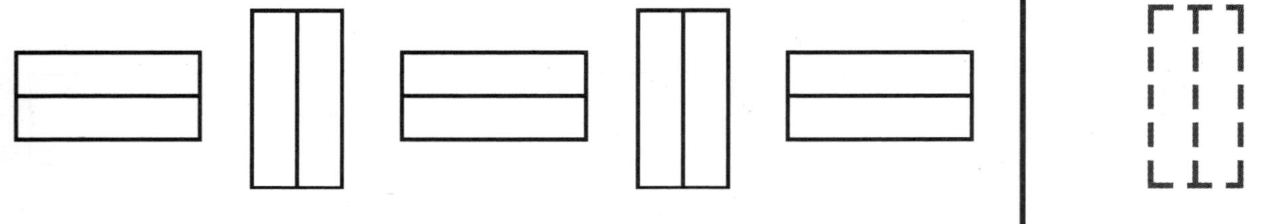

2.

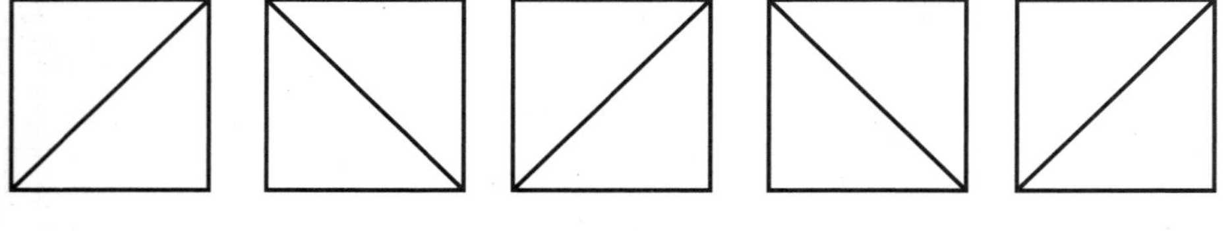

3.

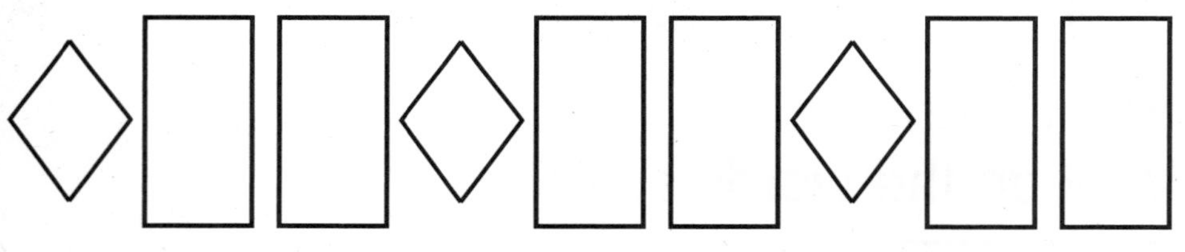

4.

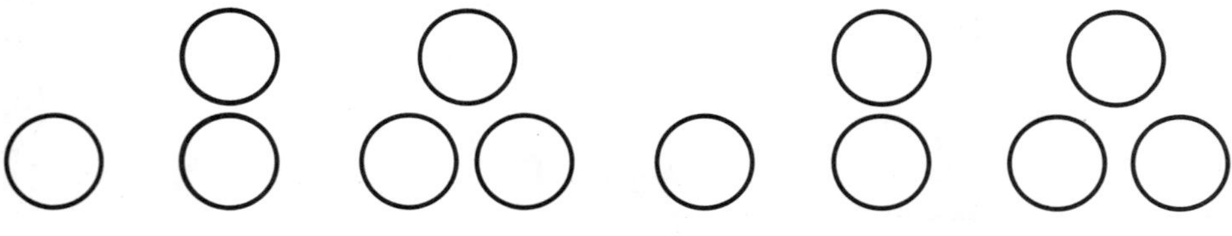

5.

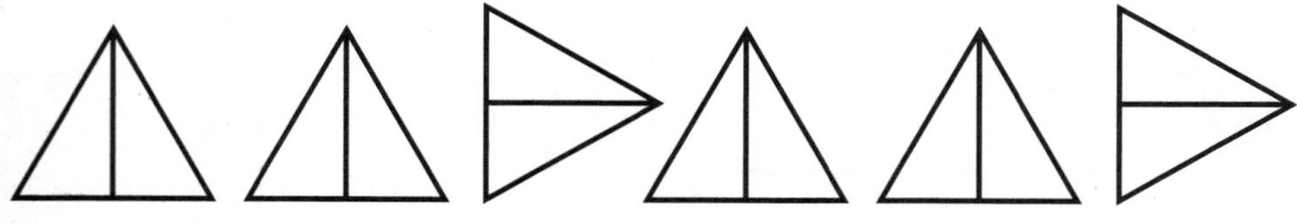

Name ___

Reproducing and Extending Patterns

Use △ and ○.
Make up your own pattern for each row.
Draw 9 shapes in your pattern.

Name ______________________________

Making and Extending Patterns

Use the shapes to make your own pattern.

1.

2.

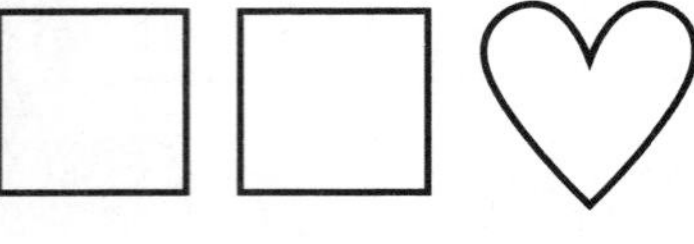

3.

4.

5.

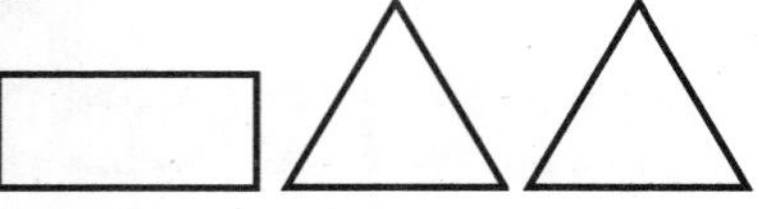

6.

Name ___

Analyzing Patterns

Follow the path.
Cross out the mistakes in the pattern.
Paste the shapes to show the pattern the correct way.

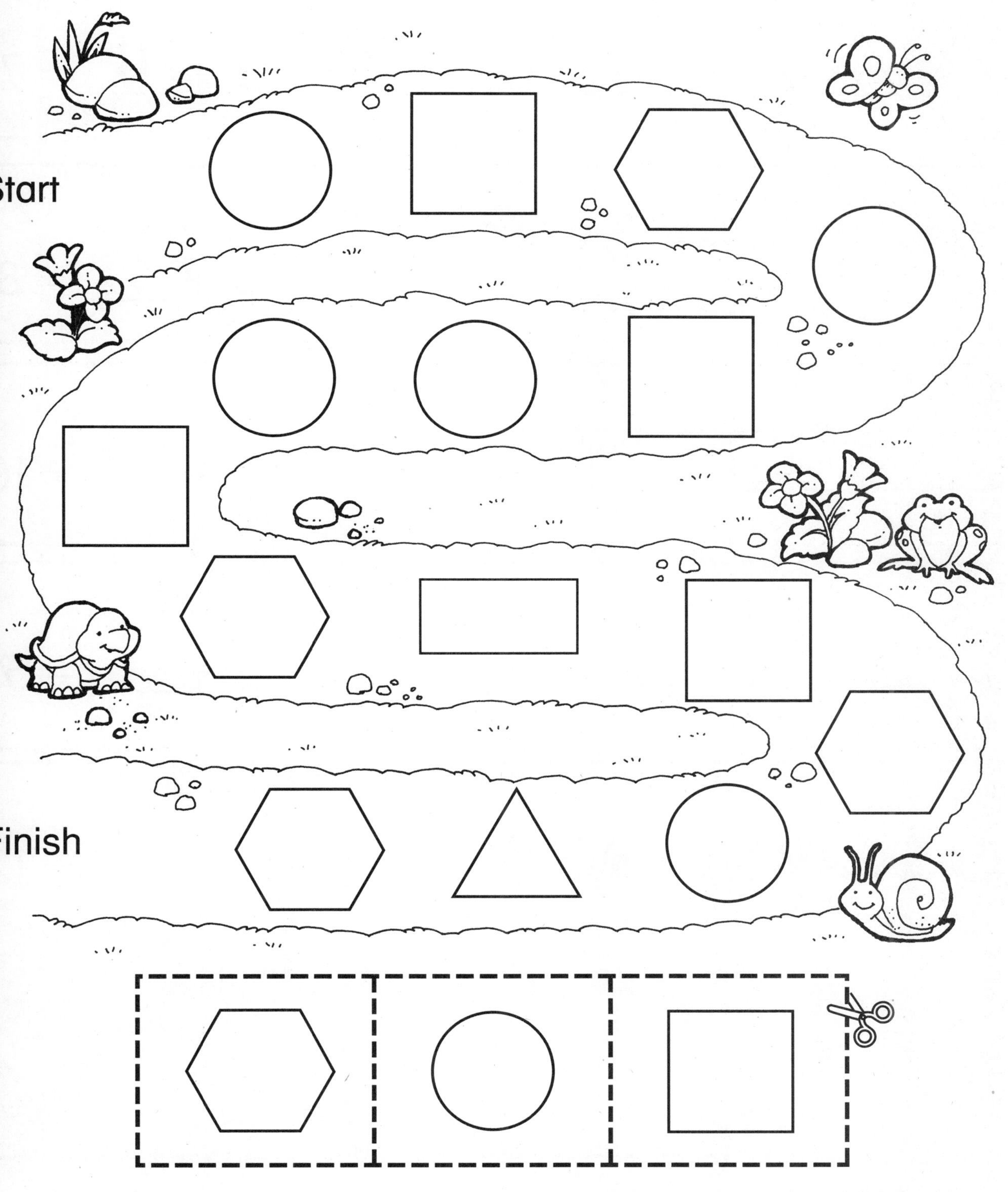

Name ______________________________

LESSON 11.1

Counting On to 12

Draw more to show the sum.
Write the missing number.

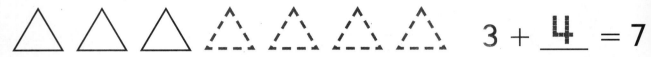

1.

3 + 4 = 7

2.

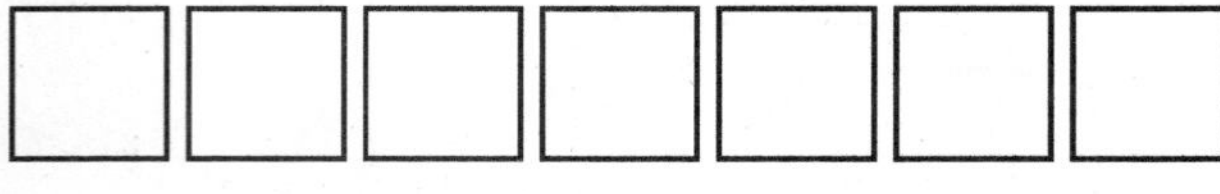

6 + ___ = 8

3.

7 + ___ = 10

4.

5 + ___ = 9

5.

9 + ___ = 12

Name ______________________________

LESSON 11.2

Doubles to 12

Solve the problem.

1. Tim picked 1 apple.
Sally gave him 1 more.
How many apples does he have in all?

2 apples

2. Tracy had 2 dresses.
Her mother got her 2 more.
How many dresses does she have in all?

_____ dresses

3. Jim saw 4 baby birds.
Then he saw 4 more.
How many birds did he see in all?

_____ birds

4. Matt found 6 eggs.
His sister found the same number.
How many eggs did they find in all?

_____ eggs

5. Mike ate 3 cookies.
Morgan ate the same number of cookies.
How many cookies did they eat in all?

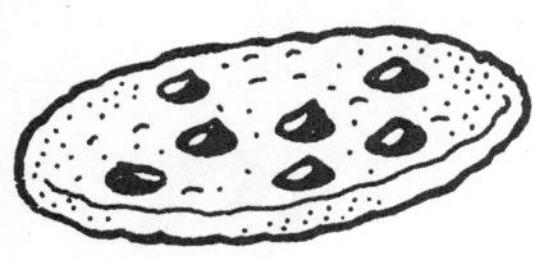

_____ cookies

6. Karen wrapped 5 presents.
Judy wrapped the same number.
How many presents did they wrap in all?

_____ presents

Name ______________________________

Three Addends

Add the prices to solve.

1. Trent has 12¢. He buys three things. Circle the three things he buys.

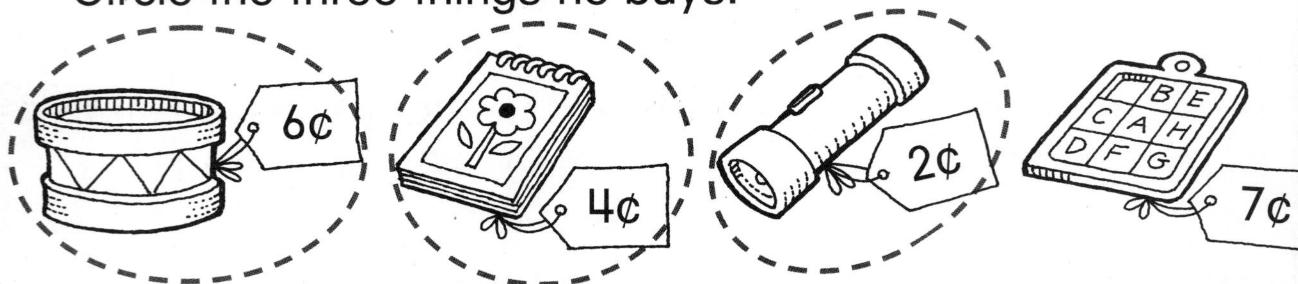

2. Iris has 8¢. She buys three things. Circle the three things she buys.

3. Abe has 11¢. He buys three things. Circle the three things he buys.

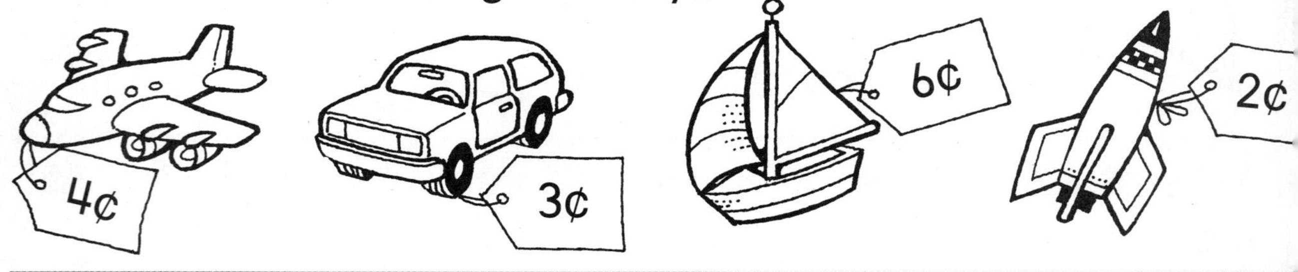

4. Jeannie has 9¢. She buys three things. Circle the three things she buys.

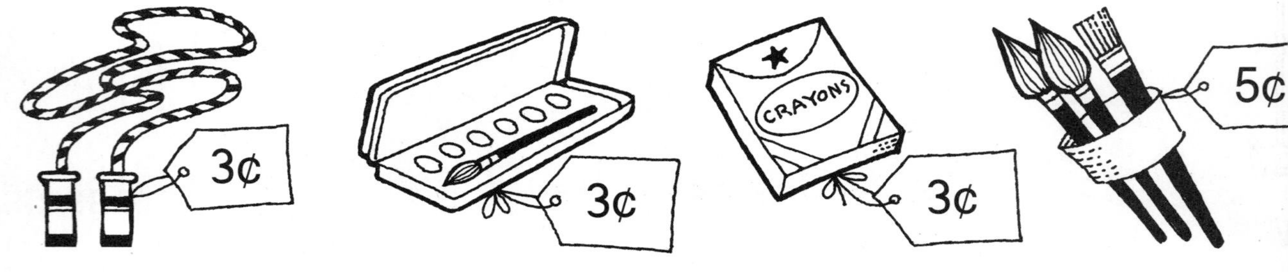

Name ______________________________

Practice the Facts

Add. Color the spaces that have the sum of 11 or 12.

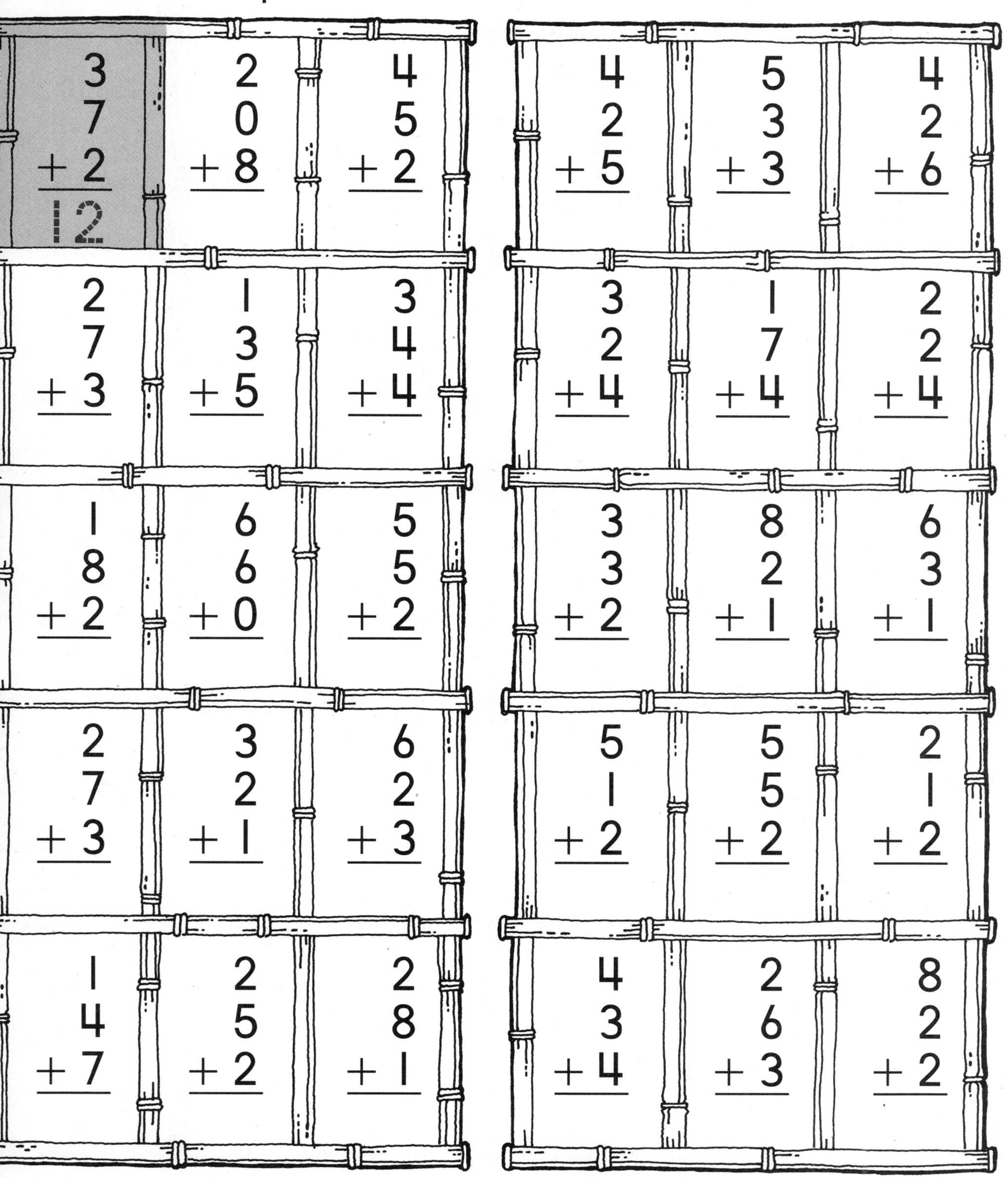

Write the word you see. ______

Name ______________________________

LESSON 11.5

Problem Solving • Act It Out

Act it out. Write the missing number.
Complete the number sentence.

1. 6 frogs were in the pond.
 3 more frogs jumped in.
 There were 9 frogs in all.

$6 + \underline{3} = 9$

2. 5 fish swam near the rocks.
 ____ fish swam by the boat.
 There were 11 fish in all.

$5 + \_\_\_\_ = 11$

3. James threw 5 balls.
 Emily threw ____ balls.
 They threw 7 balls in all.

$5 + \_\_\_\_ = 7$

4. Philip baked 6 muffins.
 Jody baked ____ muffins.
 They baked 12 muffins in all.

$6 + \_\_\_\_ = 12$

5. The hen laid 4 eggs.
 Then she laid ____ more eggs. She laid 8 eggs in all.

$4 + \_\_\_\_ = 8$

6. The rabbit hopped 1 time.
 Then it hopped ____ more times. It hopped 9 times in all.

$1 + \_\_\_\_ = 9$

Name ______________________________

Relating Addition and Subtraction

Subtract. Then write the two addition facts.

1.

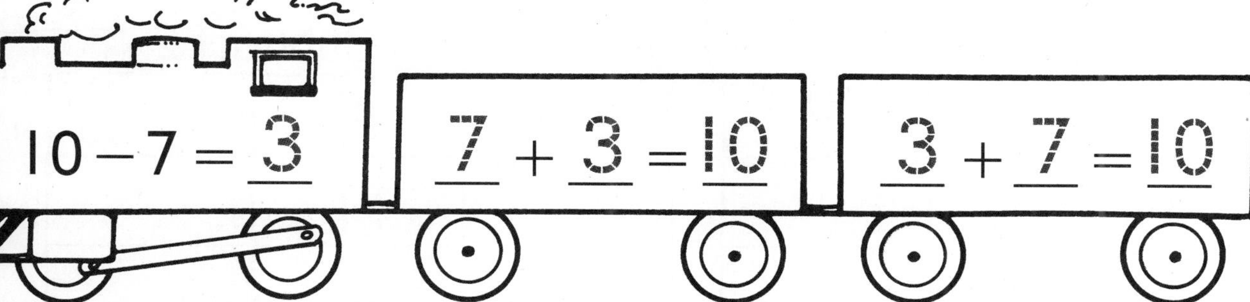

2.

3.

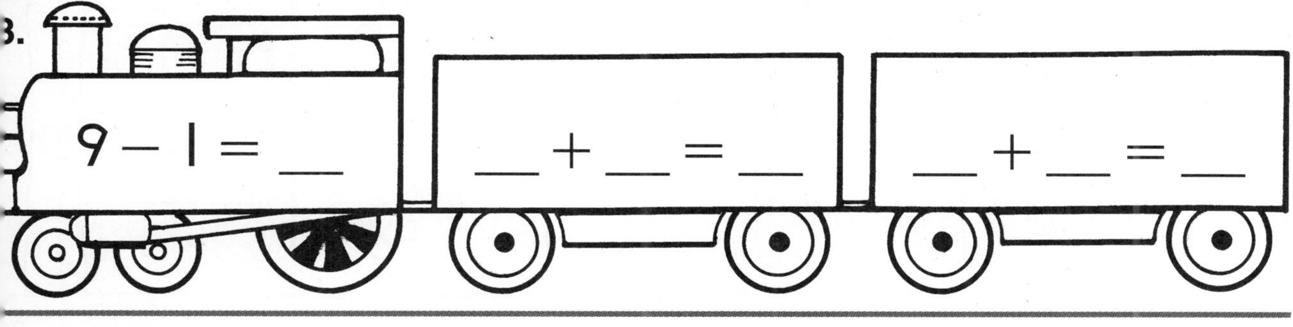

4.

5.

7 − 3 = __

__ + __ = __

__ + __ = __

Name ________________________________

Counting Back

Complete each table.

1.

Subtract 1.	
9	8
7	
10	
6	

2.

Subtract 2.	
8	
11	
6	
4	

3.

Subtract 3.	
5	
9	
3	
12	

4.

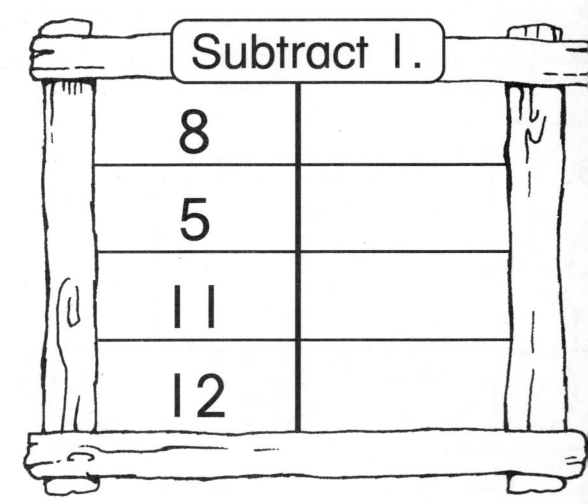

Subtract 1.	
8	
5	
11	
12	

5.

Subtract 2.	
12	
9	
5	
3	

6.

Subtract 3.	
7	
4	
10	
8	

Name ______________________________

Compare to Subtract

Count to find out how many.
Complete the table.

1.

How many?	
dog	5
rooster	
cow	
duck	
pig	
sheep	

Compare. Write the subtraction sentence.

2. How many more dogs than sheep are there?

_____ − _____ = _____

3. How many more roosters than dogs are there?

_____ − _____ = _____

4. How many fewer cows than ducks are there?

_____ − _____ = _____

5. How many fewer pigs than ducks are there?

_____ − _____ = _____

Name ____________________

Fact Families

Finish writing the fact family.
Use the three numbers in the box.

1. | 2 | 3 | 5 |

2 + 3 = 5

3 + 2 = 5

5 − 3 = 2

5 − 2 = 3

2. | 4 | 6 | 10 |

4 + 6 = 10

___ + 4 = ___

___ − ___ = 6

___ − ___ = 4

3. | 2 | 7 | 9 |

2 + 7 = 9

___ + 2 = ___

___ − ___ = 7

___ − ___ = 2

4. | 3 | 8 | 11 |

3 + 8 = 11

___ + 3 = ___

___ − ___ = 8

___ − ___ = 3

5. | 2 | 6 | 8 |

2 + 6 = 8

___ + 2 = ___

___ − ___ = 6

___ − ___ = 2

6. | 4 | 8 | 12 |

4 + 8 = 12

___ + 4 = ___

___ − ___ = 8

___ − ___ = 4

Name ____________________

Problem Solving • Write a Number Sentence

Add and subtract to solve. Write two number sentences.

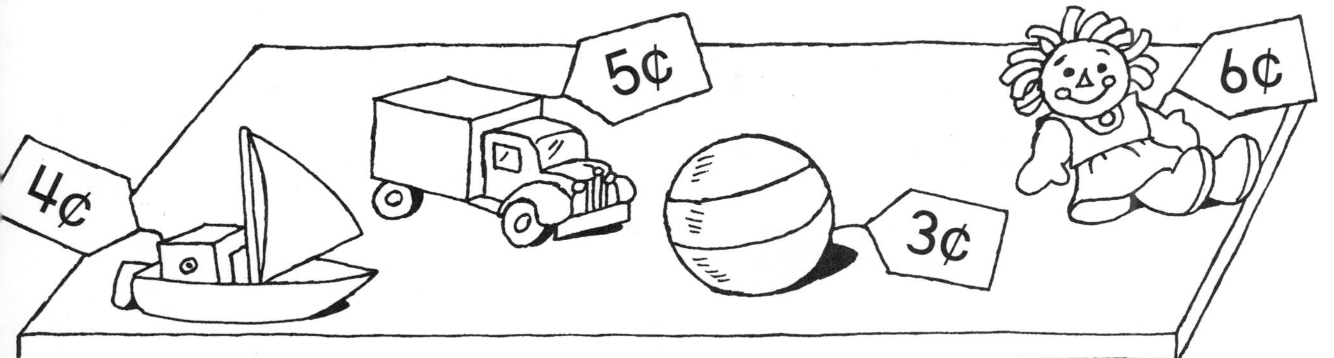

1. Andy has 12 pennies. He buys the boat and the truck. How much money does he have left?

 4 ¢ + 5 ¢ = 9 ¢

 12 ¢ − 9 ¢ = 3 ¢

 3 ¢ left

2. Patty has 10 pennies. She buys the ball and the doll. How much money does she have left?

 ____¢ + ____¢ = ____¢

 ____¢ − ____¢ = ____¢

 ____¢ left

3. Katie has 9 pennies. She buys the boat and the ball. How much money does she have left?

 ____¢ + ____¢ = ____¢

 ____¢ − ____¢ = ____¢

 ____¢ left

4. Kyle has 11 pennies. He buys the ball and the truck. How much money does he have left?

 ____¢ + ____¢ = ____¢

 ____¢ − ____¢ = ____¢

 ____¢ left

Name ______________________________

Tens

Color groups of 10 to show the number.

1. Color 30.

2. Color 60.

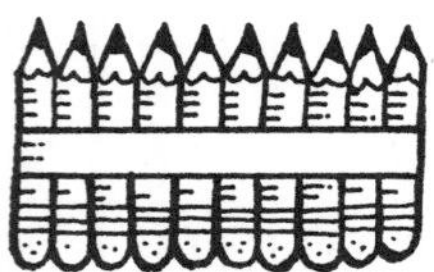
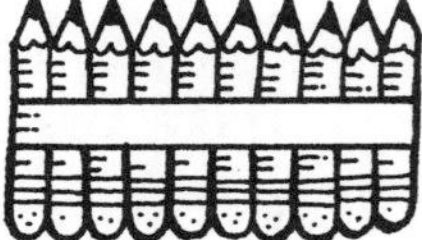
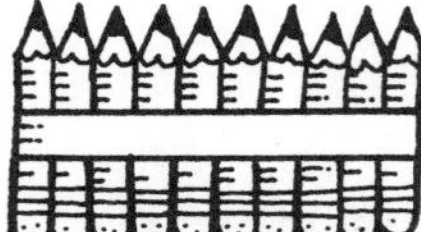
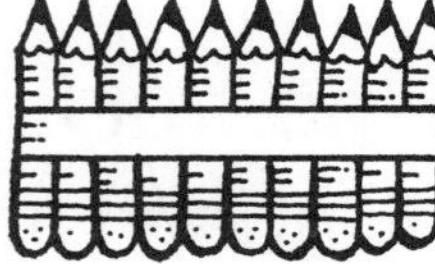
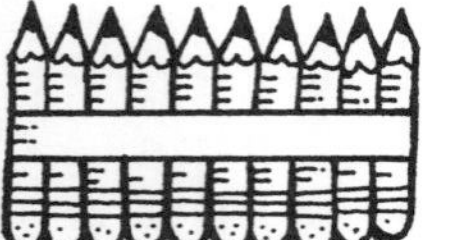
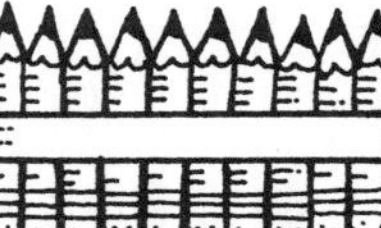
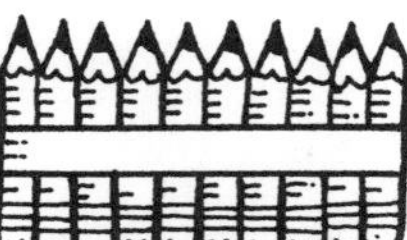
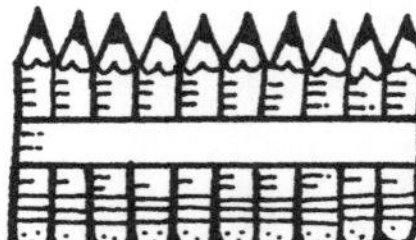
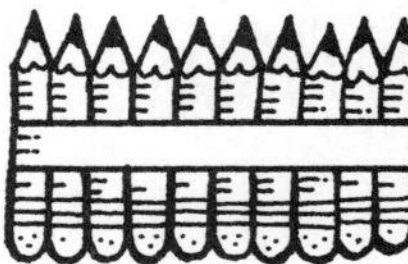

3. Color 50.

4. Color 20.

Name ______________________

Tens and Ones to 20

Read the clues.
Write the number.

1. I have 1 ten.
I have 0 ones.
What number am I?
10

2. I have 1 ten.
I have 7 ones.
What number am I?
____

3. I have 1 ten.
I have 5 ones.
What number am I?
____

4. I have 2 tens.
I have 0 ones.
What number am I?
____

5. I have 1 ten.
I have 6 ones.
What number am I?
____

6. I have 1 ten.
I have 8 ones.
What number am I?
____

7. I have 1 ten.
I have 4 ones.
What number am I?
____

8. I have 1 ten.
I have 1 one.
What number am I?
____

Name ______________________________

Tens and Ones to 50

Find the row of 24 books. Color the books red.
Find the row of 45 books. Color them blue.
Color the row of 17 books yellow.
Color the row of 36 books green.
Color the row of 52 books orange.

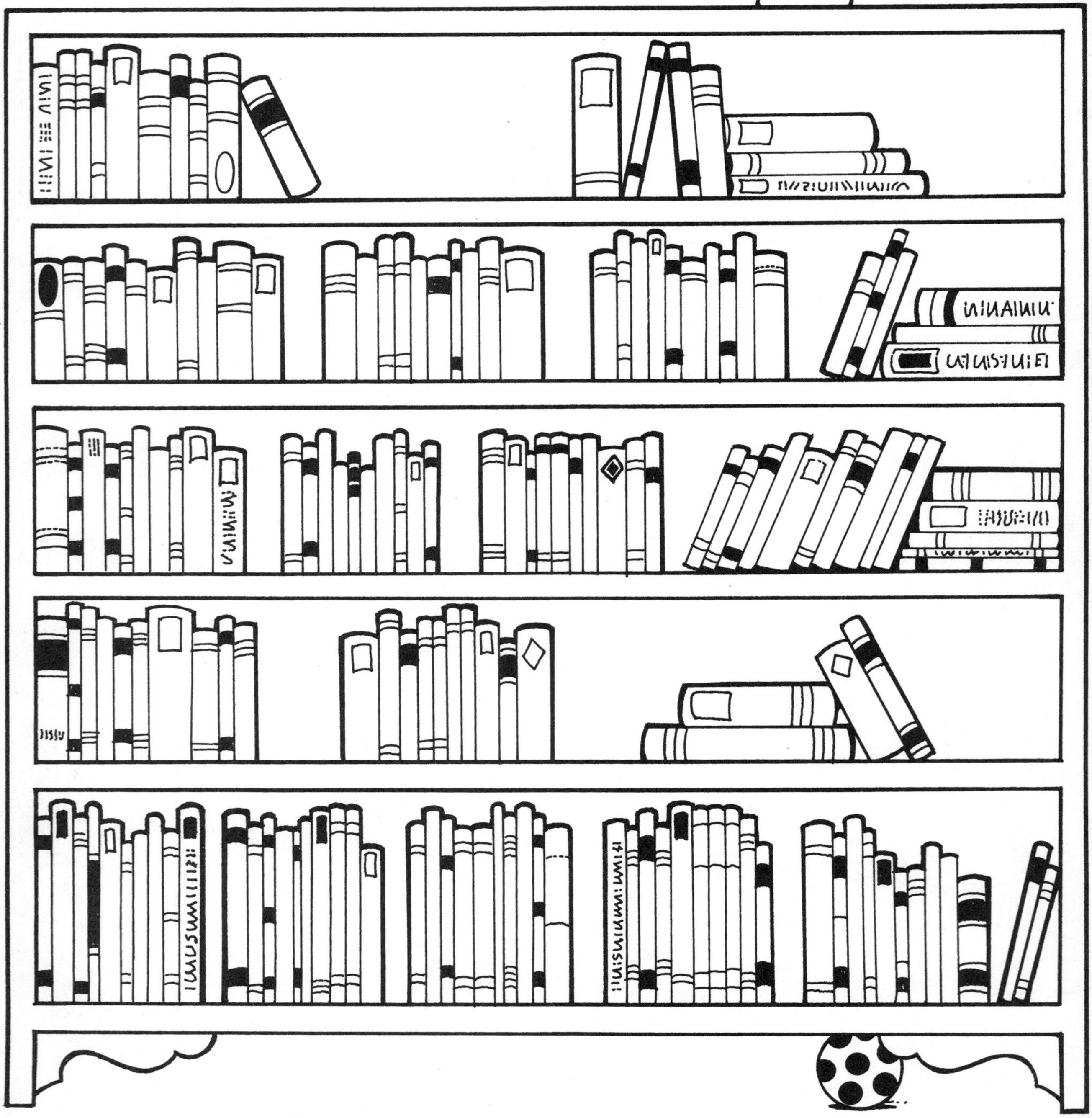

Name ____________________

LESSON 13.4

Tens and Ones to 80

Look for the numbers. Color the picture.

7 in the tens place blue

8 in the tens place green

5 in the tens place red

6 in the ones place yellow

2 in the ones place black

3 in the ones place brown

Name ______________________________

Tens and Ones to 100

Use what you know about pennies and dimes to solve the problem.

10 pennies = 1 dime

1. Andy has 7 dimes. He finds 3 pennies. How much money does he have?

   73 ¢

2. Kelli has 6 dimes. She finds 2 pennies. How much money does she have?

   ______ ¢

3. Ernie has 4 dimes. He finds 7 pennies. How much money does he have?

   ______ ¢

4. Cindy has 5 dimes. She finds 1 penny. How much money does she have?

   ______ ¢

5. Roxie has 3 dimes. She finds 8 pennies. How much money does she have?

   ______ ¢

6. Wayne has 2 dimes. He finds 6 pennies. How much money does he have?

   ______ ¢

Name ______________________

LESSON 13.6

Estimating 10

Estimate how many stars.
Circle your estimate.

1.

between 50 and 60   between 70 and 80

2.

between 20 and 30   between 30 and 40

3.

between 60 and 70   between 80 and 90

4.

between 30 and 40   between 40 and 50

Name ____________________

LESSON 14.1

Ordinals

Write a word from the box to answer the problem.

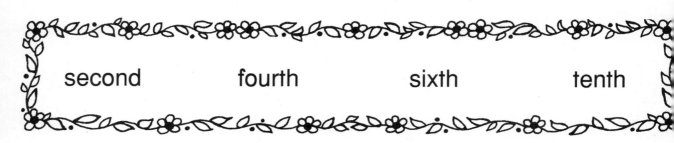

second	fourth	sixth	tenth

1. Amy is first in line.
Jack is third in line.
Paul is standing between them.
In which place is Paul?

second

2. Katie was third in line.
She left the line.
Matt was fifth in line.
In which place is Matt now?

____

3. There are 10 children in line. Tasha is the next to the last in line. Rico is standing behind her.
In which place is Rico?

____

4. Allen is fourth in line.
Callie is next in line.
Then comes Emma.
In which place is Emma?

____

Name ____________________

LESSON 14.2

Greater Than

1. Color each that has a number greater than 30.

2. Color each that has a number greater than 60.

3. Color each that has a number greater than 90.

4. Color each that has a number greater than 50.

Name ____________________

Less Than

1. Show a number that is 1 less.

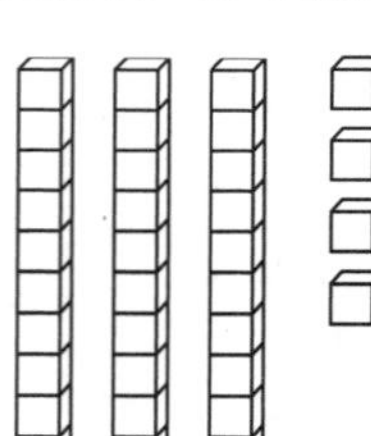

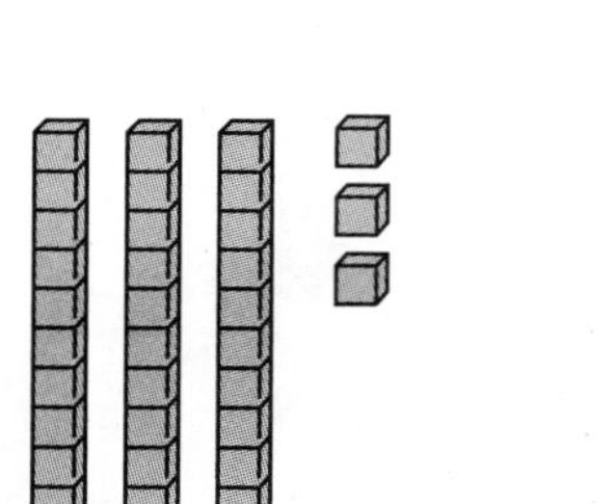

2. Show a number that is 10 less.

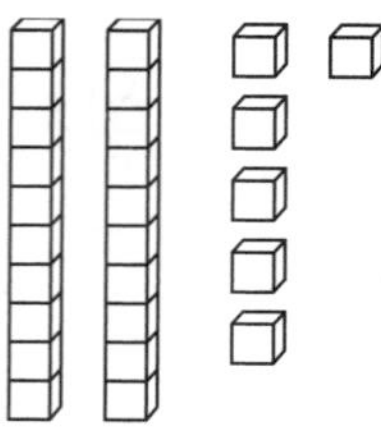

3. Show a number that is 4 less.

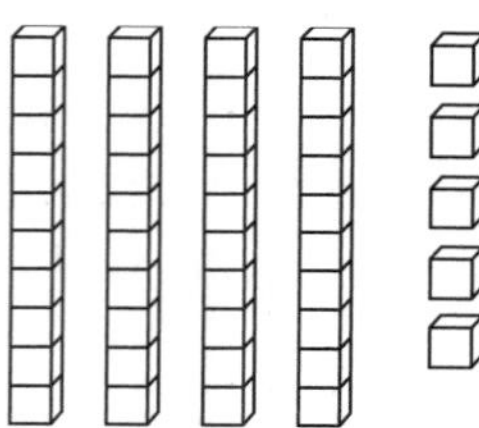

4. Show a number that is 10 less.

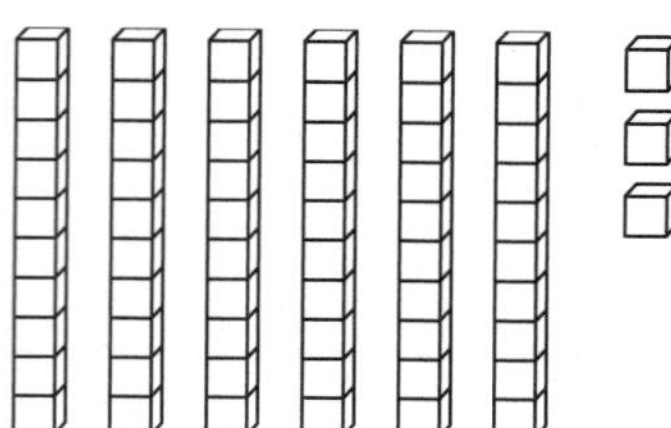

5. Show a number that is 6 less.

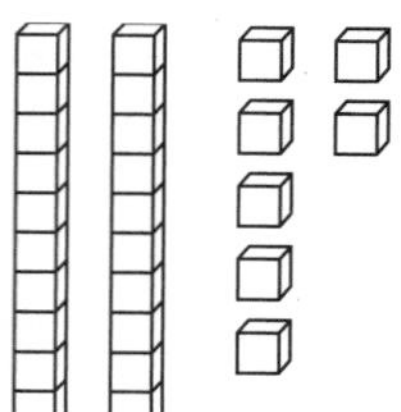

Name ______________________________

LESSON 14.4

Before, After, Between

Read the riddle. Write the number.

1. I am a number between 40 and 42. What number am I?

   41

2. I am a number that comes after 29 and before 31. What number am I?

   ______

3. I am a number between 16 and 18. What number am I?

   ______

4. I am a number that comes after 65 and before 67. What number am I?

   ______

5. We are numbers that come after 23 and before 27. What numbers are we?

   ______ ______ ______

6. We are numbers that come after 44 and before 48. What numbers are we?

   ______ ______ ______

7. We are numbers that come after 28 and before 32. What numbers are we?

   ______ ______ ______

8. We are numbers that come after 50 and before 54. What numbers are we?

   ______ ______ ______

Name ______________________________

LESSON 14.5

Order to 100

Write the missing numbers.

1.

2.

3.

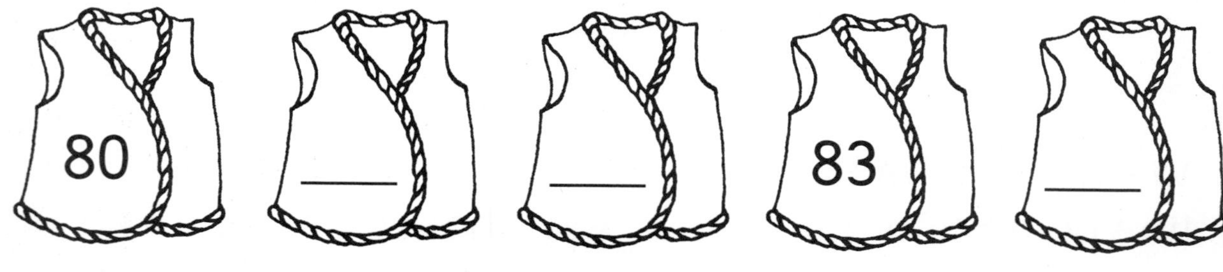

4.

5.

Name ___________________________

Counting by Tens

Count by tens to make a path for the mouse to get to the cheese.

10	21	32	28	29
20	15	40	39	27
18	30	43	50	44
34	58	62	60	65
75	71	70	81	76
77	69	68	80	85
87	74	83	90	92
54	96	82	98	100

Name ___

Counting by Fives

Count by fives to drive the truck to the market.

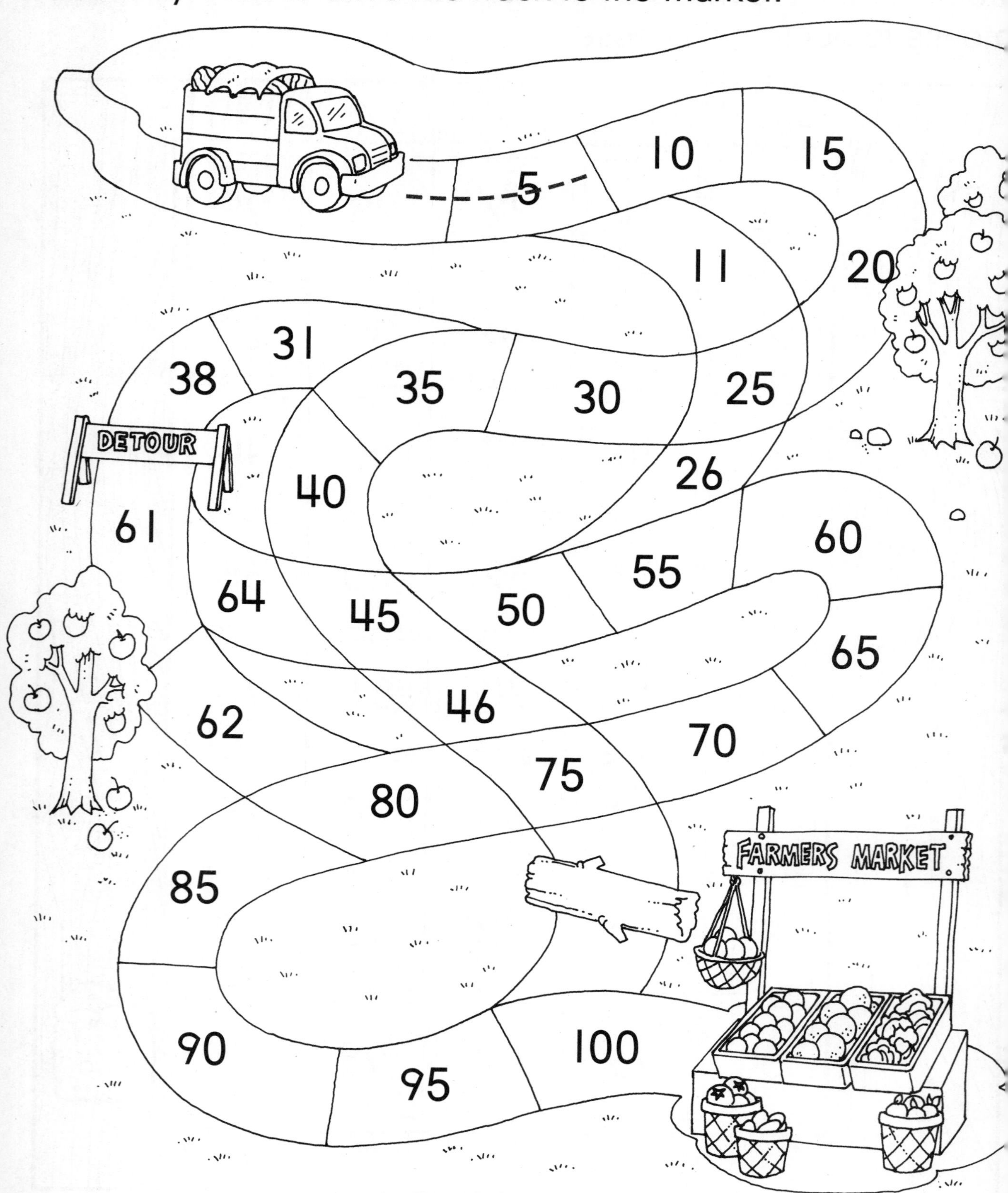

ame ______________________________

Counting by Twos

Cut and paste. Count by twos
o match the bee to the flower.

1.

2.

3.

4.

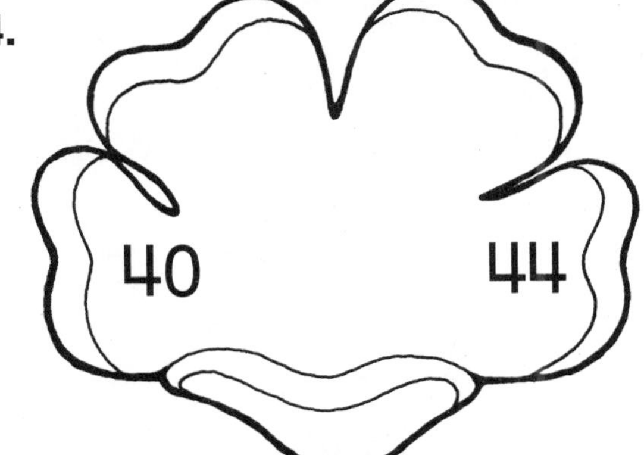

5.

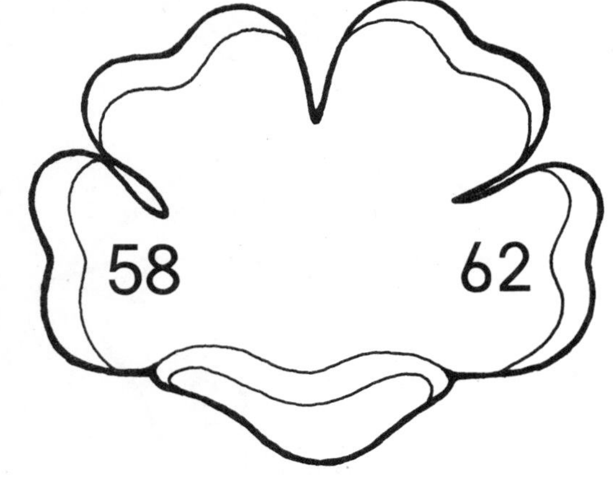

6.

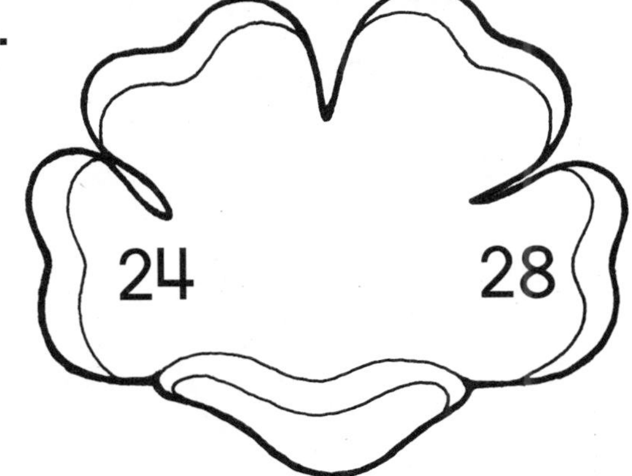

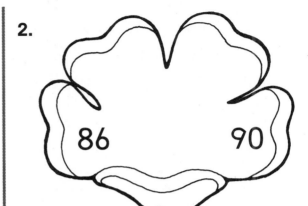

Name ____________________

Even and Odd Numbers

Solve each riddle.

1. We are odd numbers.
We are less than 10
but greater than 3.
What numbers are we?

5, 7, 9

2. We are even numbers.
We are less than 13
but greater than 8.
What numbers are we?

____ ____

3. We are odd numbers.
We are less than 40
but greater than 36.
What numbers are we?

____ ____

4. We are even numbers.
We are less than 20
but greater than 12.
What numbers are we?

____ ____ ____

Color even numbers yellow.
Color odd numbers green.

5.

6.

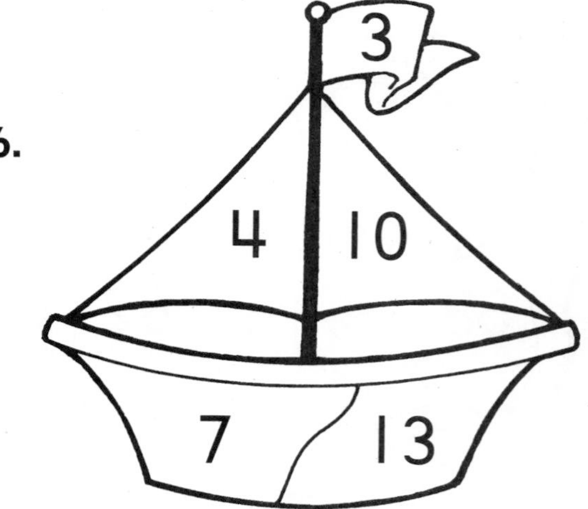

Name ______________________________

Let's Go Shopping!

Cut and paste pennies and nickels to show how much each food costs.

1.

2.

3.

4.

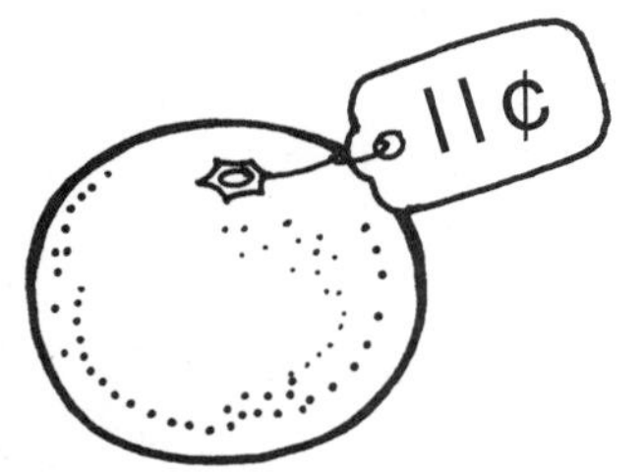

Name ______________________________

Toy Town

Draw the number of dimes you need to buy each toy.

1. 

10¢ 10¢ 10¢ 10¢ 10¢

2. 

3. 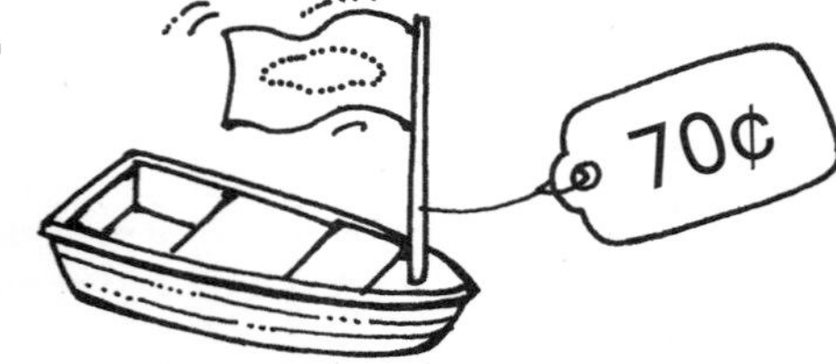

4. 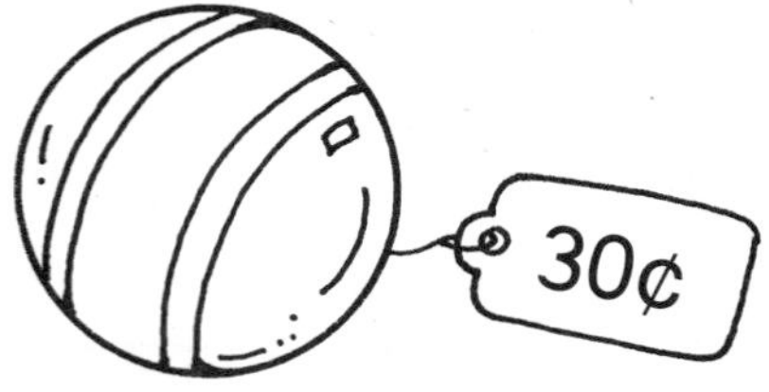

5. 

ame ___________________________

LESSON 16.3

How Much Money?

Use pennies and nickels to solve each problem.

1. Paco has 1 nickel. He saves 3 nickels. How much money does Paco have now? 20 ¢	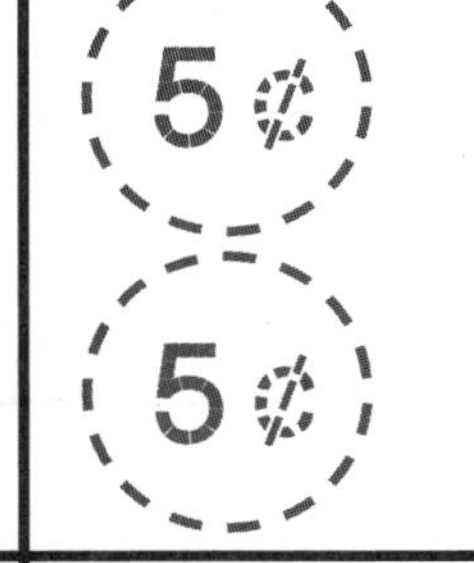 
2. Lee has 5 pennies and 5 nickels. How much money does Lee have in all? ____ ¢	
3. Gwen has 4 nickels in her piggy bank. Her mother gives her 1 nickel and 3 pennies. How much money does Gwen have now? ____ ¢	
4. Carlos has 7 pennies in his pocket. He has 7 nickels in his hand. How much money does Carlos have in all? ____ ¢	

Name ____________________

Coin Purse Collections

Cut and paste pennies and dimes to show how much money is in each purse.

Name ______________________________

LESSON 16.5

At the Fair

You have [4 dimes, 2 nickels, 4 pennies].

Color the things you would buy.

1. How many things did you buy?

______

2. How much money did you spend?

______¢

3. How much money do you have left?

______¢

Name ____________________

LESSON 17.1

It's in the Bank!

You have these coins.	Trade for the fewest coins. Draw the coins.
1. 10 	10¢
2. 35 	
3. 20 	
4. 30 	
5. 45 	

Name ______________________

Equal Amounts

Show the amount in two ways.
Circle the way that uses fewer coins.

1.

	dimes	nickels	pennies
15¢	1	1	
15¢		2	5

2.

	dimes	nickels	pennies
30¢			
30¢			

3.

	dimes	nickels	pennies
28¢			
28¢			

4.

	dimes	nickels	pennies
25¢			
25¢			

5.

	dimes	nickels	pennies
35¢			
35¢			

Name ______________________

LESSON 17.3

How Much Is Needed?

You want to buy a snack. Use the fewest coins to buy each food. Write the numbers.

	 Dime	 Nickel	 Penny
1.  18¢	1	1	3
2.  26¢			
3.  47¢			
4.  35¢			
5.  9¢			

Name ______________________________

Quarter

1. Mack has one coin that equals 25¢.
Circle the coin.

2. Now he has 3 coins that equal 25¢. Circle the coins.

3. Now Mack has 4 coins that make 25¢. Circle the coins.

4. Now he has 5 coins that make 25¢. Circle the coins.

5. Now he has 7 coins that make 25¢. Circle the coins.

Name ______________________

Problem Solving • Act It Out

Go shopping. Use coins to solve.

You have these coins.	You want to buy these things.	Use the fewest coins to show the money you need. Draw the coins.
1. 	 15¢	1¢ 1¢
2. 	 45¢	
3. 	 42¢	
4. 	 36¢	

Name ______________________________

LESSON 18.1

Ordering Months and Days

January

S	M	T	W	T	F	S
			1	2	3	4
5	6	7	8	9	10	11
12	13	14	15	16	17	18
19	20	21	22	23	24	25
26	27	28	29	30	31	

February

S	M	T	W	T	F	S
						1
2	3	4	5	6	7	8
9	10	11	12	13	14	15
16	17	18	19	20	21	22
23	24	25	26	27	28	

March

S	M	T	W	T	F	S
						1
2	3	4	5	6	7	8
9	10	11	12	13	14	15
16	17	18	19	20	21	22
23/30	24/31	25	26	27	28	29

April

S	M	T	W	T	F	S
		1	2	3	4	5
6	7	8	9	10	11	12
13	14	15	16	17	18	19
20	21	22	23	24	25	26
27	28	29	30			

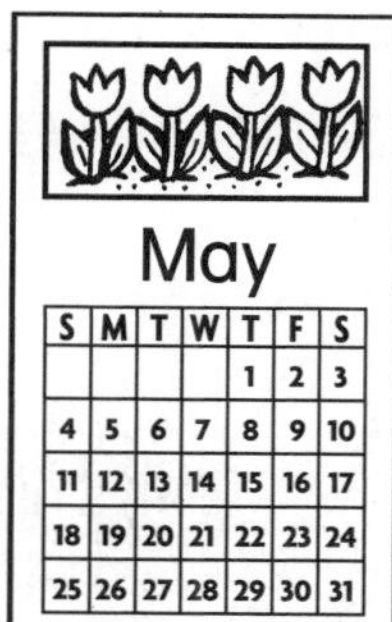
May

S	M	T	W	T	F	S
				1	2	3
4	5	6	7	8	9	10
11	12	13	14	15	16	17
18	19	20	21	22	23	24
25	26	27	28	29	30	31

June

S	M	T	W	T	F	S
1	2	3	4	5	6	7
8	9	10	11	12	13	14
15	16	17	18	19	20	21
22	23	24	25	26	27	28
29	30					

July

S	M	T	W	T	F	S
		1	2	3	4	5
6	7	8	9	10	11	12
13	14	15	16	17	18	19
20	21	22	23	24	25	26
27	28	29	30	31		

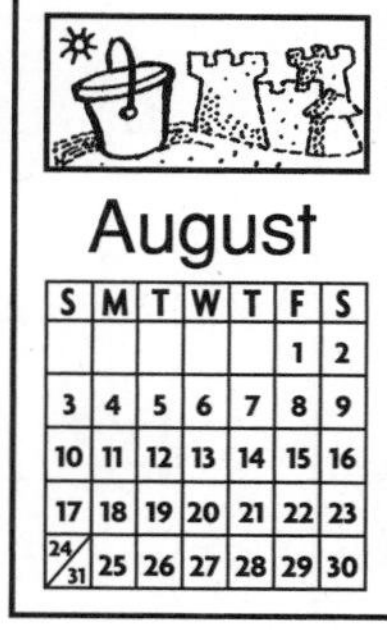
August

S	M	T	W	T	F	S
					1	2
3	4	5	6	7	8	9
10	11	12	13	14	15	16
17	18	19	20	21	22	23
24/31	25	26	27	28	29	30

September

S	M	T	W	T	F	S
	1	2	3	4	5	6
7	8	9	10	11	12	13
14	15	16	17	18	19	20
21	22	23	24	25	26	27
28	29	30				

October

S	M	T	W	T	F	S
			1	2	3	4
5	6	7	8	9	10	11
12	13	14	15	16	17	18
19	20	21	22	23	24	25
26	27	28	29	30	31	

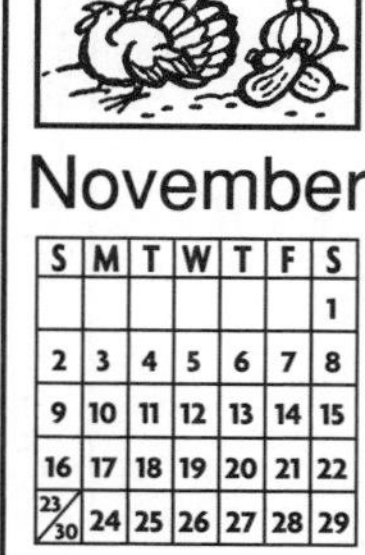
November

S	M	T	W	T	F	S
						1
2	3	4	5	6	7	8
9	10	11	12	13	14	15
16	17	18	19	20	21	22
23/30	24	25	26	27	28	29

December

S	M	T	W	T	F	S
	1	2	3	4	5	6
7	8	9	10	11	12	13
14	15	16	17	18	19	20
21	22	23	24	25	26	27
28	29	30	31			

Use the calendar.

1. I am the first month of the year. What month am I? ______________________

2. I am the month that comes after November. ______________________

3. I am the third month of the year. ______________________

4. I am the month that comes after June. ______________________

5. I am your birthday month. ______________________

Name ______________________________

LESSON 18.2

Reading the Calendar

Find the dates of the birthdays.
Use the calendar to answer the questions.

November

Sunday	Monday	Tuesday	Wednesday	Thursday	Friday	Saturday
	1 Leon	2	3	4	5	6
7	8	9	10 Sonia	11	12	13
14	15	16	17	18	19	20
21	22 Maria	23	24	25	26	27 Gary
28	29	30 Kay				

1. What is the date of Kay's birthday? ______
2. What is the date of Gary's birthday? ______
3. What is the date of Maria's birthday? ______
4. Who has a birthday 9 days after Leon? ______________

Name ______________________________

LESSON 18.3

Ordering Events

Draw what will happen next.

Name ______________________________

LESSON 18.4

Estimating Time

Number these things in the order of how long they take to do. Write **1** to show the thing that takes the shortest time. Number **6** will show the thing that takes the longest. If you have a clock with a second hand, check your answers.

Activity	Order
Tell a funny story.	
Write the letters A to Z.	
Sing a song.	
Take three breaths.	
Jump 10 times.	
Wash your hands.	

Name ________________________________

LESSON 19.1

Reading the Clock

My School Day			
9:00	Reading	1:00	Science
10:00	Spelling	2:00	Social Studies
11:00	Math	3:00	Go home
12:00	Lunch		

Circle the clock that shows the correct time.

1. Eat lunch.

2. Take a spelling test.

3. Work in a math book.

4. Go home.

Name ____________________

Hour

Draw the hour hand so that both clocks show the same time.

1.

10:00

2.

6:00

3.

2:00

4.

5:00

5.

8:00

6.

1:00

Name ______________________________

Hour

Show the time you might do each thing.
Draw the hour hand and the minute hand.

1. Get up in the morning.

2. Go to school.

3. Play with friends.

4. Watch a TV show.

5. Do homework.

6. Go to bed.

Name ____________________

LESSON 19.4

Half Hour

Look at the clock.
Write the time it will be thirty minutes later.

1.

Thirty minutes later

8:30

2.

Thirty minutes later

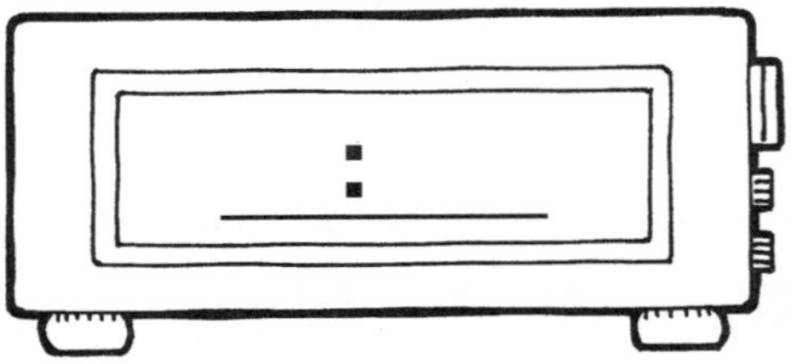

___:___

3.

Thirty minutes later

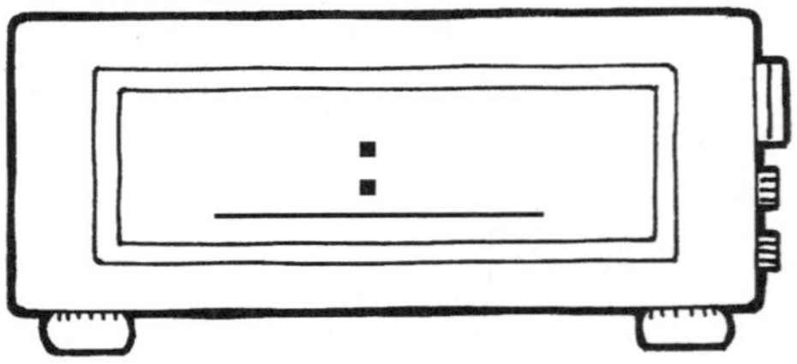

___:___

4.

Thirty minutes later

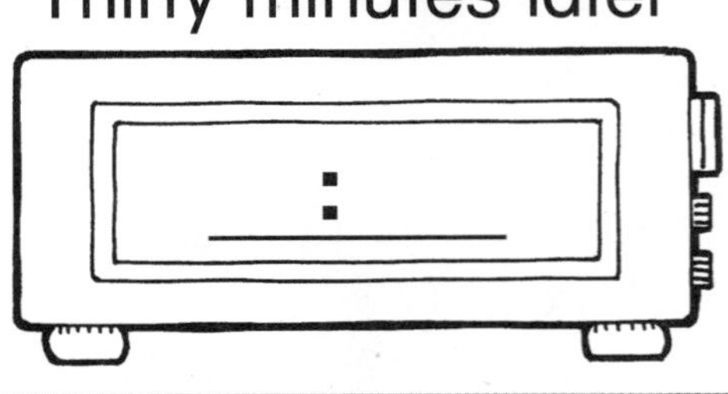

___:___

5.

Thirty minutes later

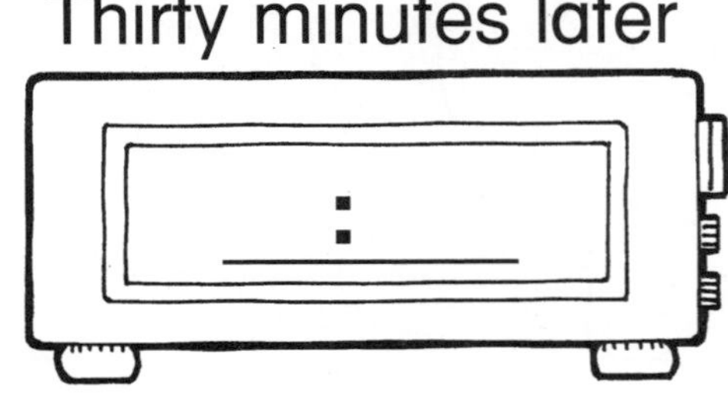

___:___

Name ____________________

Problem Solving • Act It Out

Estimate how many minutes it will take.
Then act it out to see if you are correct.
Use a clock to time how long it takes.

1. Build a block tower.

Estimate

_____ minutes

My time

_____ minutes

2. Draw a picture of your family.

Estimate

_____ minutes

My time

_____ minutes

3. Do 10 sit-ups.

Estimate

_____ minutes

My time

_____ minutes

Name ______________________________

LESSON 20.1

Using Nonstandard Units

1. Meg's doll shoe is longer than Jill's.
Estimate how many [paper clip] long each shoe is.
Use [paper clip] to measure. Color Meg's shoe [blue crayon].
Color Jill's shoe [red crayon].

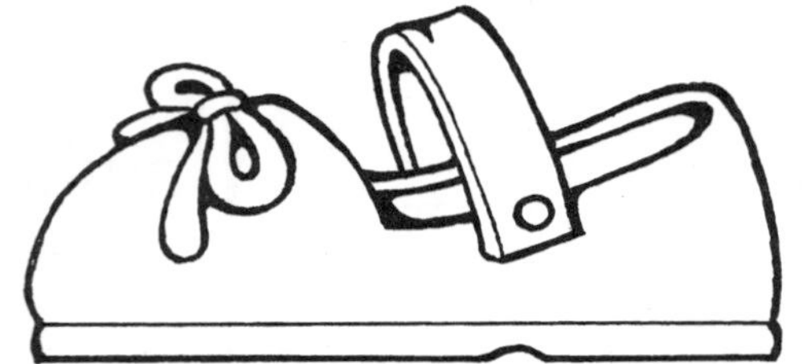

Estimate	about ____ [paper clip]	Estimate	about ____ [paper clip]
Measure	about ____ [paper clip]	Measure	about ____ [paper clip]

2. Sam's toy car is shorter than David's.
Estimate how many [cube] long each car is.
Use [cube] to measure. Color Sam's car [green crayon].
Color David's car [yellow crayon].

Estimate	about ____ [cube]	Estimate	about ____ [cube]
Measure	about ____ [cube]	Measure	about ____ [cube]

Name ______________________________

LESSON 20.2

Path into Space

Help the astronaut get to the right planet.
Estimate which path is about 12 inches long.
Use an inch ruler to measure.
Color the correct path red.

Name ___

Measuring Me

Estimate. Then use an inch ruler to measure.

	Estimate	Measure
my foot	about ___ inches	about ___ inches
my hand	about ___ inches	about ___ inches
my arm	about ___ inches	about ___ inches
my big step	about ___ inches	about ___ inches

Name ________________________________

LESSON 20.4

Measuring in Centimeter Units

1. Draw a path between two dots that is about 6 centimeters long.

2. Draw a path between two dots that is about 8 centimeters long.

3. Draw a path between two dots that is longer than 4 centimeters but shorter than 5 centimeters.

4. Draw a path between two dots that is longer than 10 centimeters but shorter than 11 centimeters.

Name ______________________________

Go for the Stars

Find these objects in your classroom. Estimate. Do you think the object will be more or less than 10 centimeters long? Color one star to show. Then use a centimeter ruler to measure. Write how many centimeters.

Object	More Than 10 Centimeters	Less Than 10 Centimeters	Centimeters
1.	☆	☆	____
2.	☆	☆	____
3.	☆	☆	____
4. ERASER	☆	☆	____
5.	☆	☆	____

Name ____________________

LESSON 21.1

Using a Balance

Find something lighter than .

Find something heavier than .

Find something that is about the same as .

Draw one thing in each box.

1.

Crayons

lighter

2.

Crayons

heavier

3.

Crayons

same

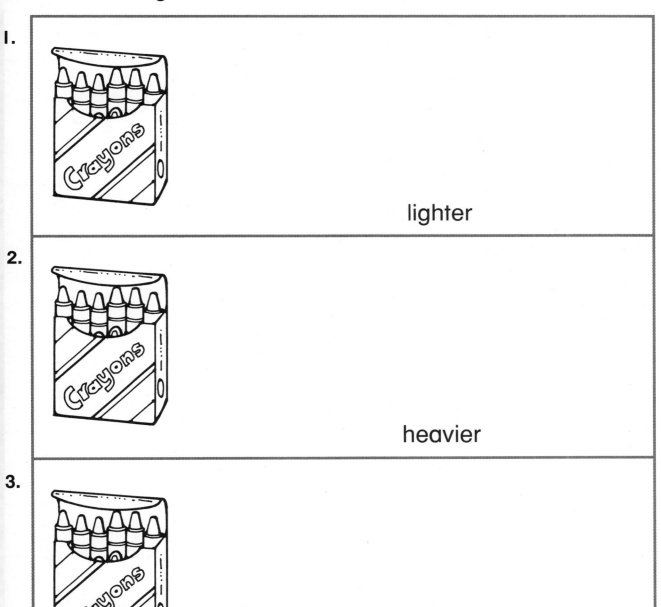

Name ____________________

Using Nonstandard Units

Circle the cup you think is lighter.
Then fill one cup with each kind of thing.
Put the two cups on a 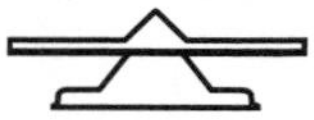 .
Which is lighter? Color that picture blue.

1.

pennies

cubes

2.

paper clips

stones

3.

cotton balls

sand

4.

marbles

chalk

Name ______________________________

Measuring with Cups

Measure how much rice each container will hold.
Then number the containers in order.
Make the one that holds the smallest amount number 1.

Name ______________________________

Temperature Hot and Cold

It is summer.
Mark an **X** on the pictures that do not belong.

PARK

Name ___

Equal and Unequal Parts of Wholes

The first figure shows equal parts.
Show equal parts a different way.

1.

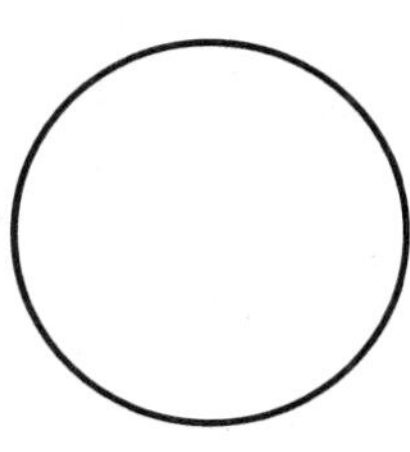

2.

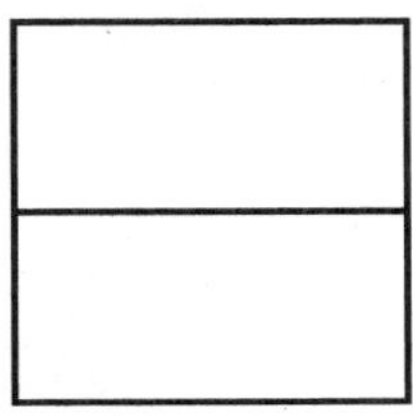

3.

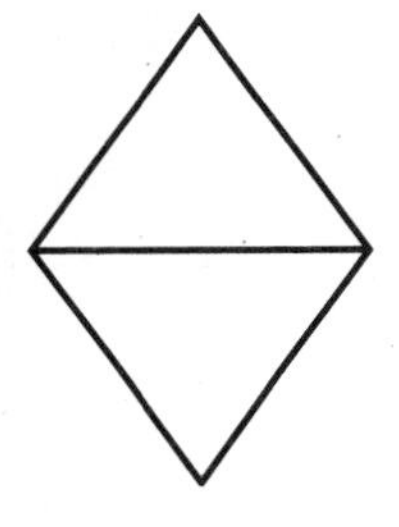

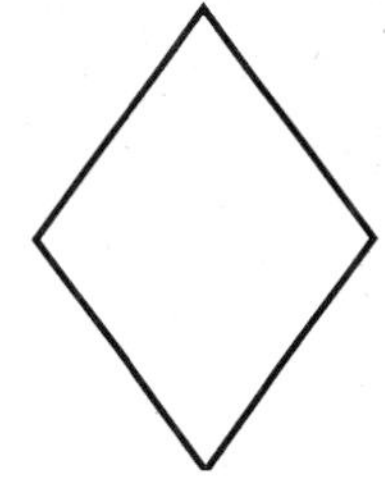

4.

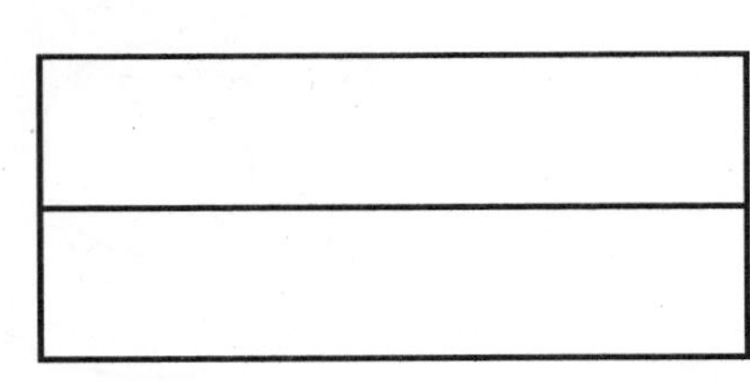

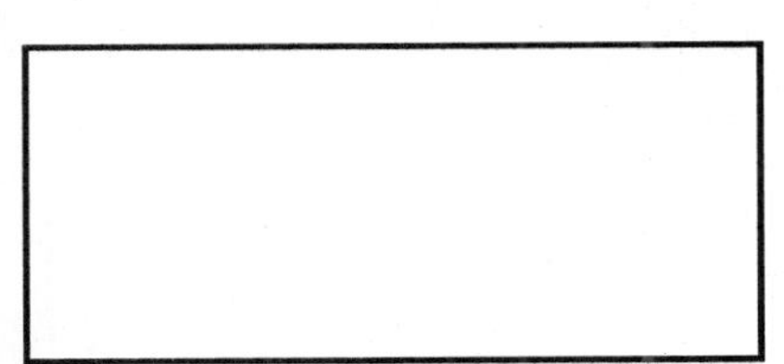

5.

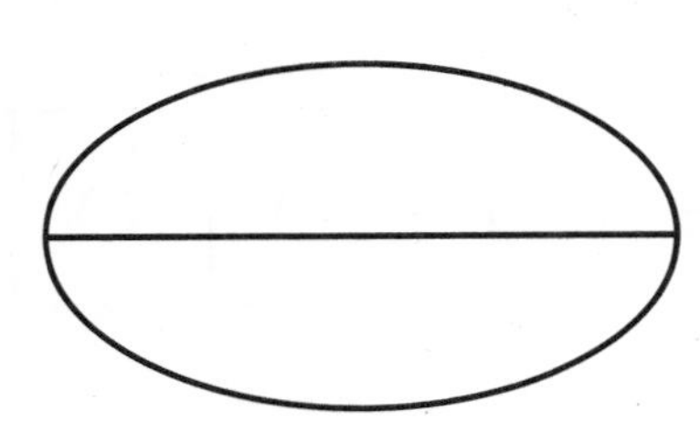

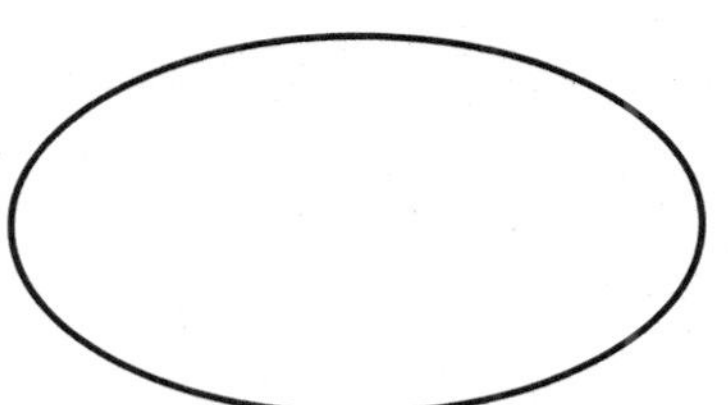

Name ____________________

Halves

Circle the pictures that show halves.
Circle their letters in the chart.
Then write the circled letters in order.
How good is your work?

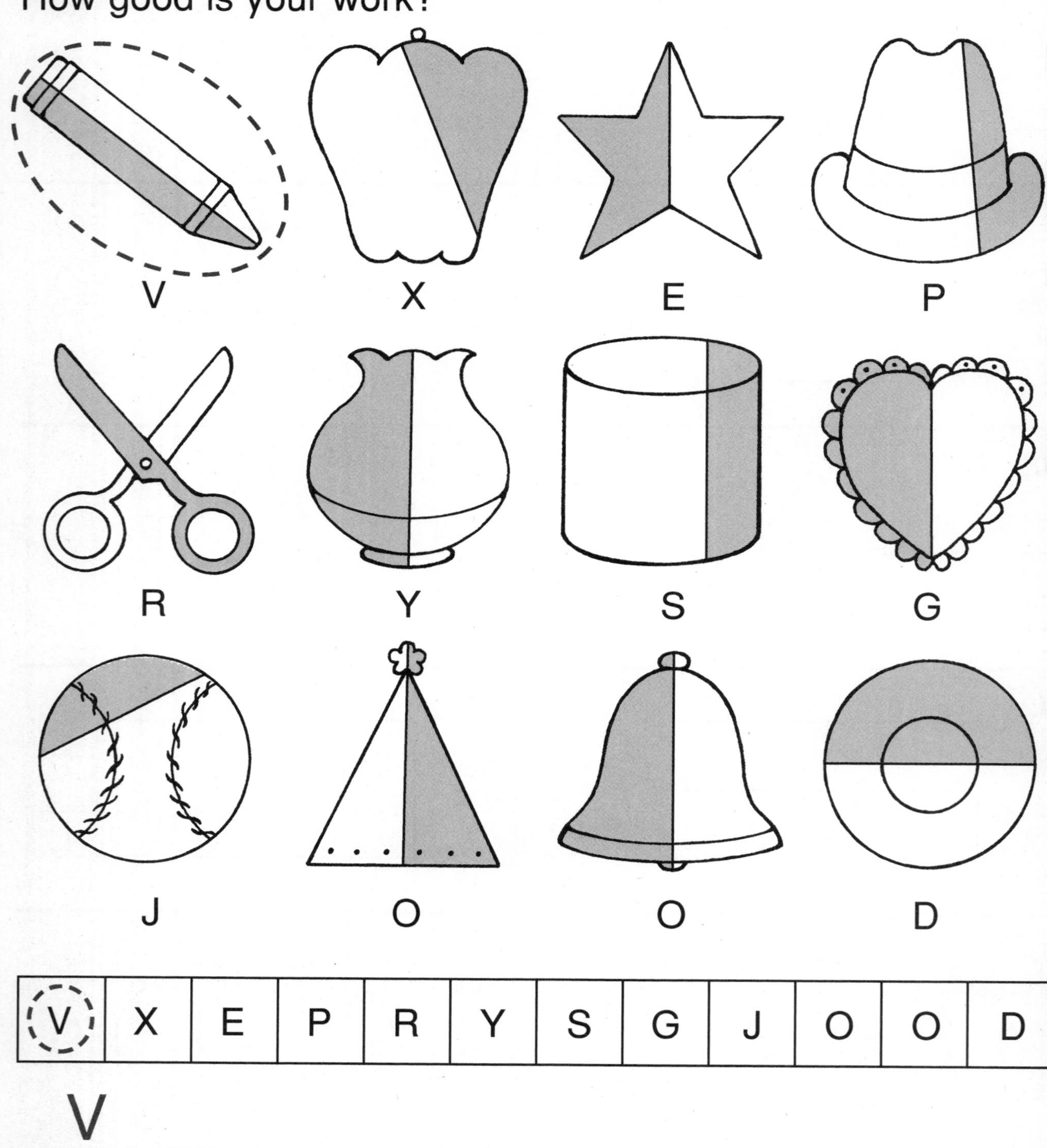

V	X	E	P	R	Y	S	G	J	O	O	D

V ___ ___ ___ ___ ___ ___ ___ ___

ame ___________________________________

ourths

ind the figures that show fourths.
hen join those figures to make a path
ome from school.

Name ______________________

Thirds

Draw lines to show equal parts.
Then color the figure to show the fraction.

1.

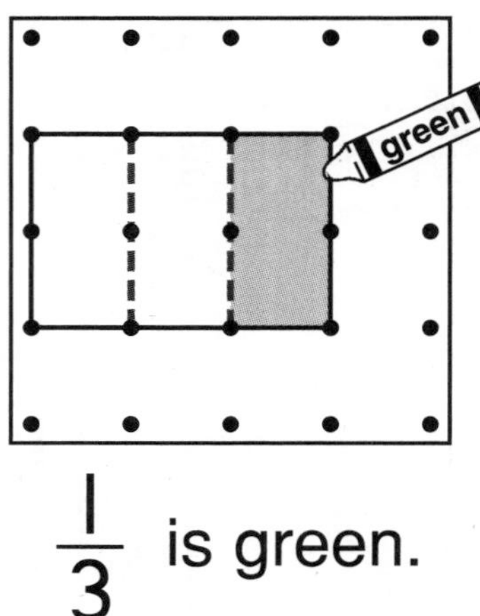

$\frac{1}{3}$ is green.

2.

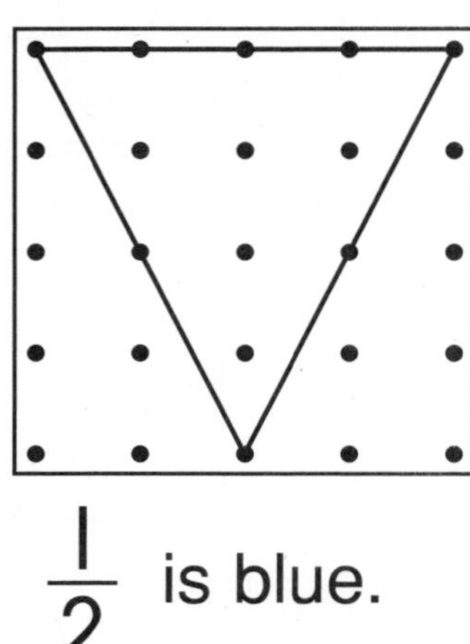

$\frac{1}{2}$ is blue.

3.

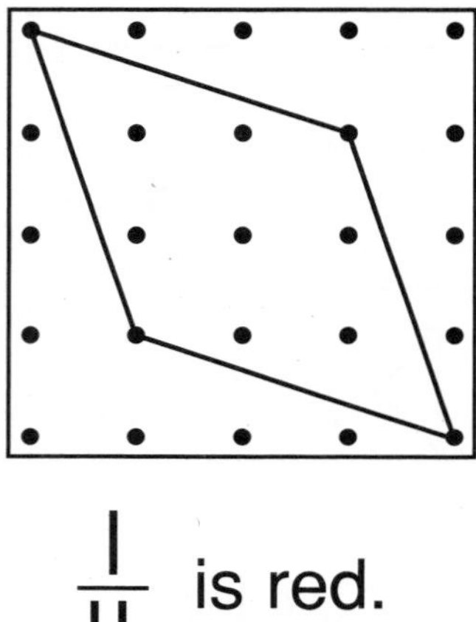

$\frac{1}{4}$ is red.

4.

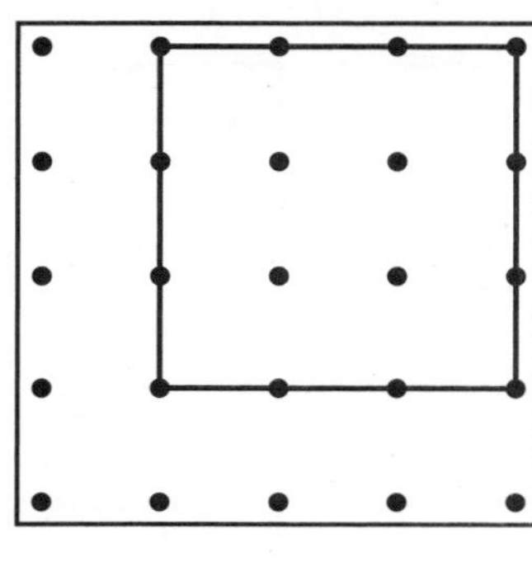

$\frac{1}{3}$ is yellow.

5.

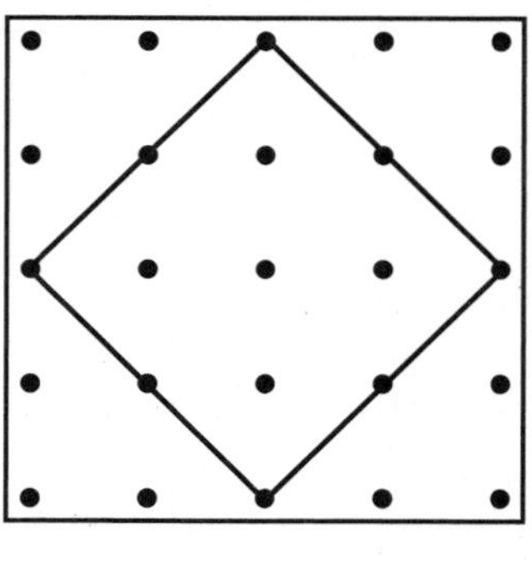

$\frac{1}{2}$ is purple.

6.

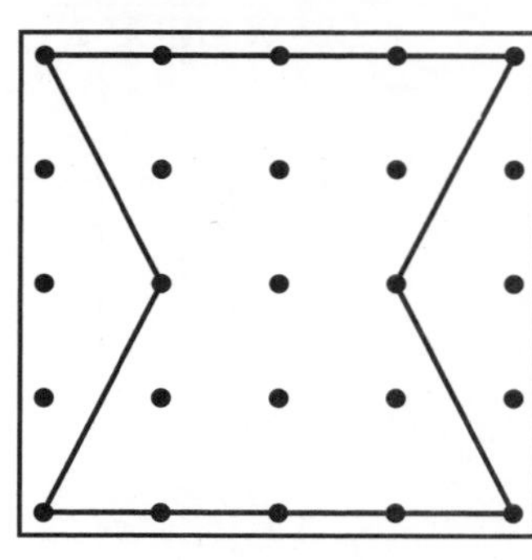

$\frac{1}{4}$ is brown.

Name ______________________________

Visualizing Results

1. Look for 2 equal parts in the triangle.
Trace lines with a crayon to show the parts.

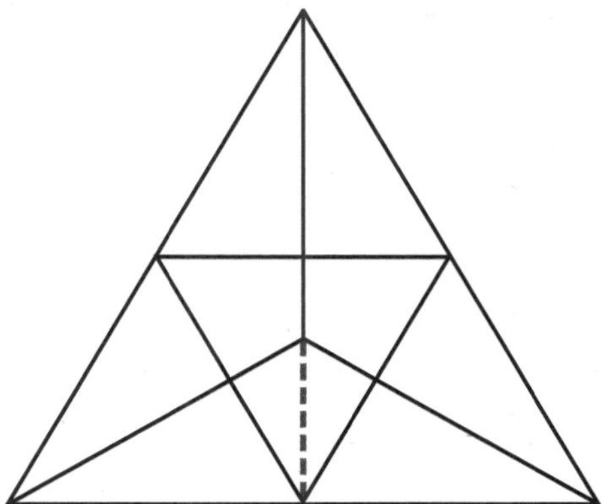

2. Look for 3 equal parts in the triangle.
Trace lines with a crayon to show the parts.

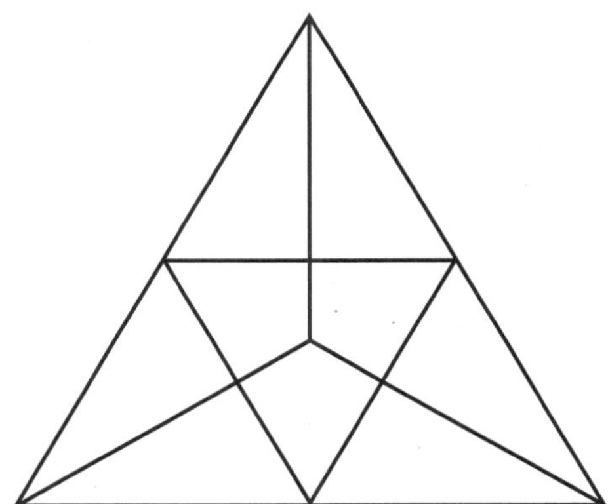

3. Look for 4 equal parts in the triangle.
Trace lines with a crayon to show the parts.

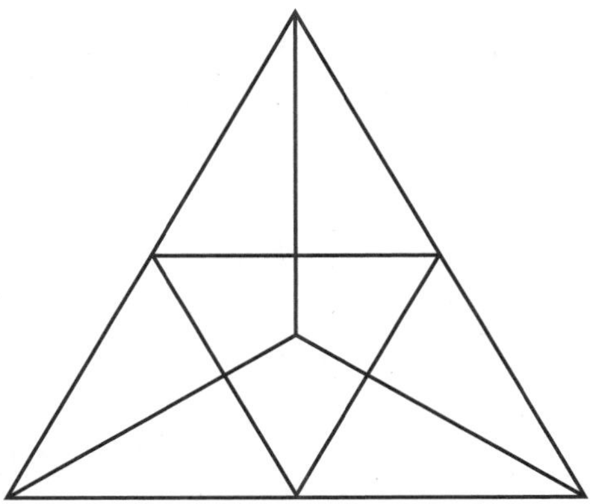

Name ______________________________

LESSON 22.6

Parts of Groups

Use the picture. Circle the fraction.

1. Alexis has 6 books. She has read 3 of them. What fraction of the books has she read?

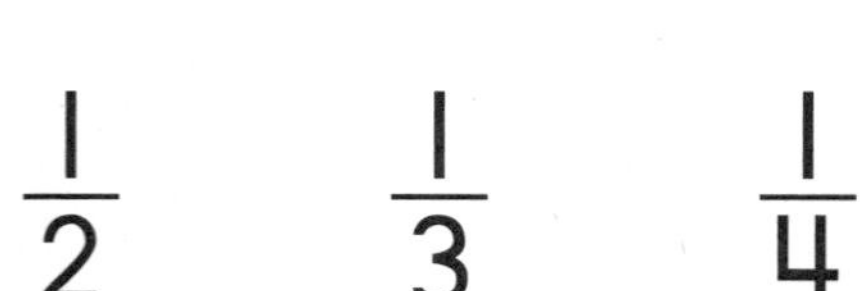

$\frac{1}{2}$ $\frac{1}{3}$ $\frac{1}{4}$

2. Rachel took 8 shirts from the dryer. Two of the shirts have a happy face. What fraction of the shirts have a happy face?

$\frac{1}{2}$ $\frac{1}{3}$ $\frac{1}{4}$

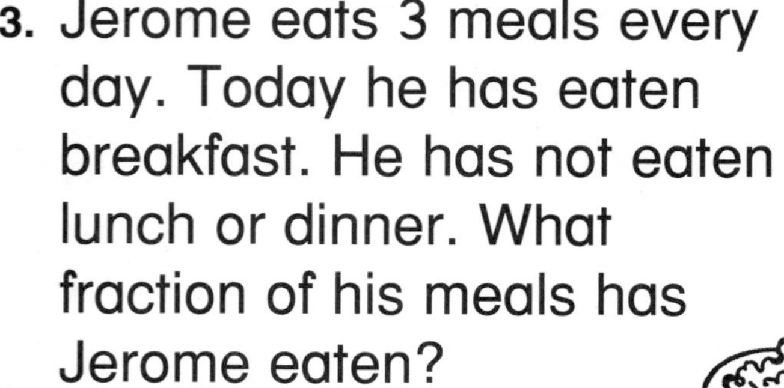

3. Jerome eats 3 meals every day. Today he has eaten breakfast. He has not eaten lunch or dinner. What fraction of his meals has Jerome eaten?

$\frac{1}{2}$ $\frac{1}{3}$ $\frac{1}{4}$

Name ________________________________

The Marching Band

Sort the children. Make a table.
Use tally marks to show how many in each group.

Sort the children another way.

Write a question about one table.
Give the question to a friend.

____

____

Name ____________________

Certain or Impossible

Draw things that can come out of this bag.
Then draw things that cannot come out of it.

Certain **Impossible**

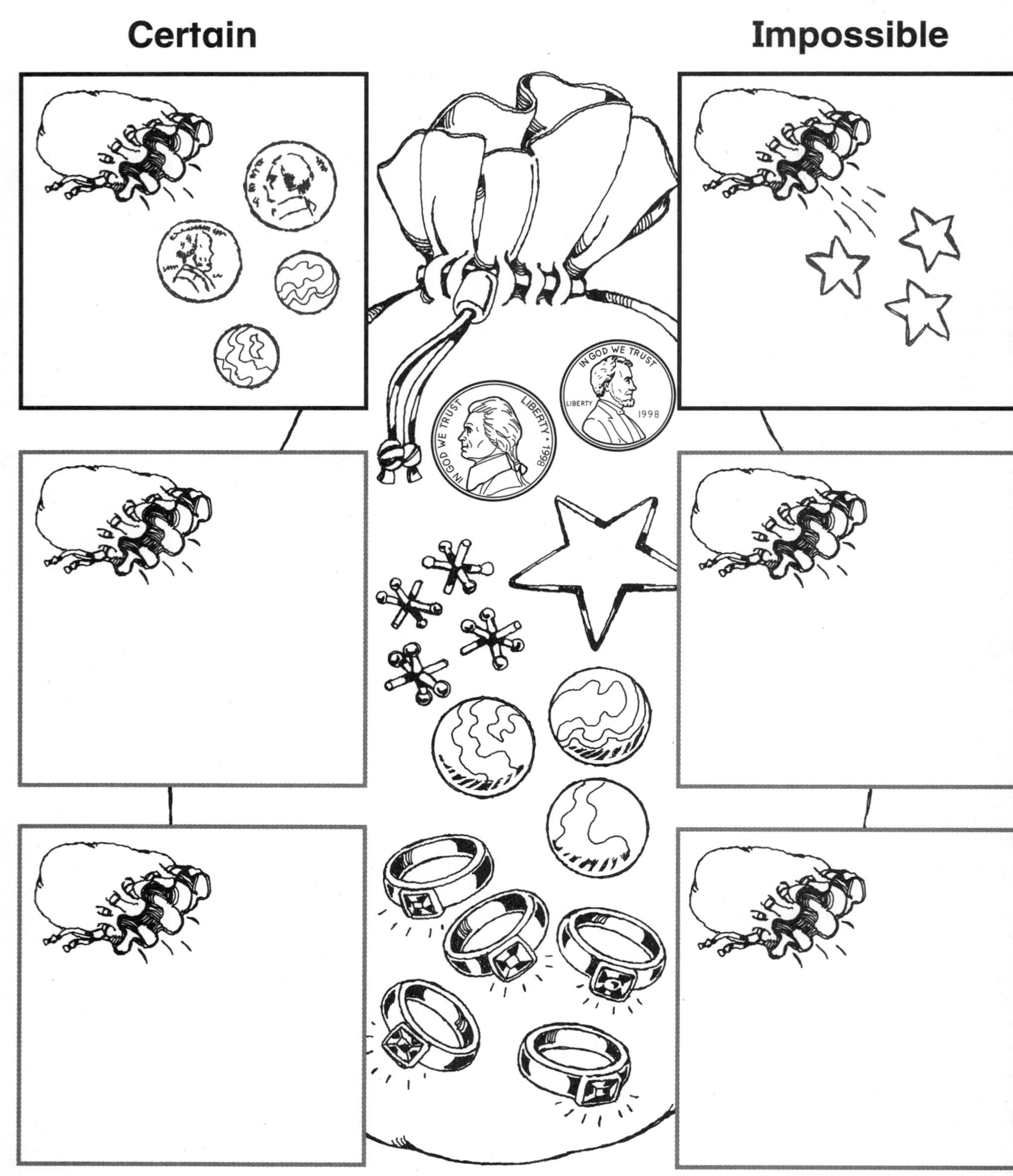

ame ___________________________________

Most Likely

Circle your prediction for picking red.

1.

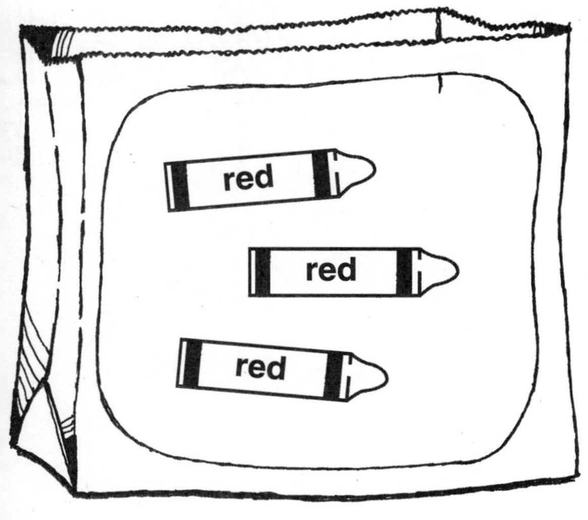

certain likely impossible

2.

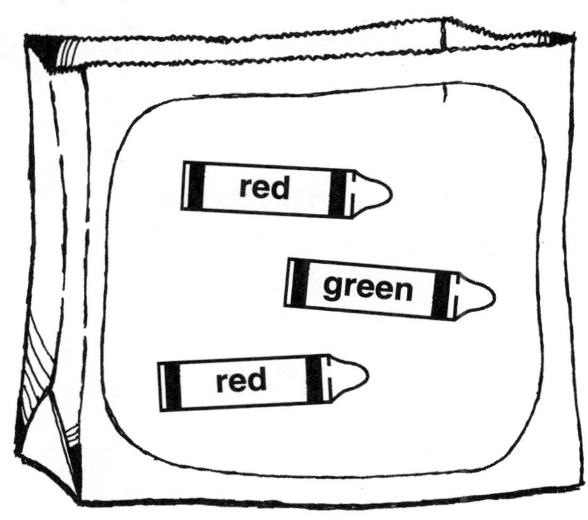

certain likely impossible

3.

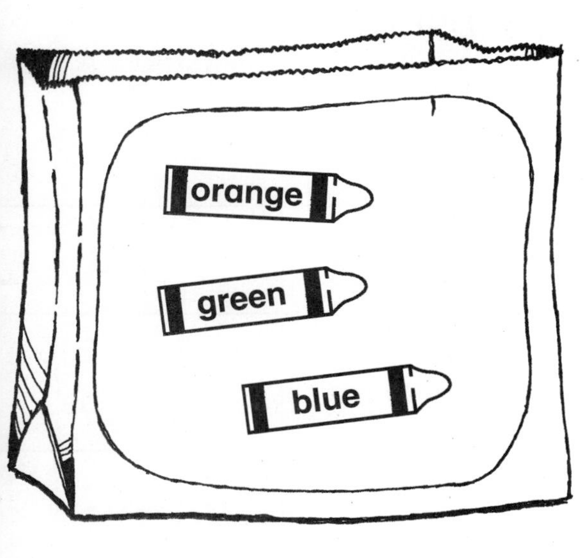

certain likely impossible

4.

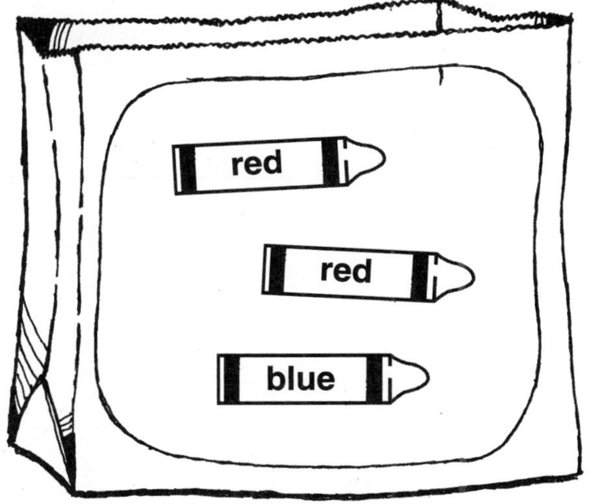

certain likely impossible

Name ______________________________

Tallying Events

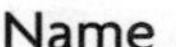

Work with a partner.
Use a bag, 5 red cubes, 5 blue cubes, and 5 green cubes.

Put the cubes into the bag. Take out one cube.

Color the cubes in the table. Make a tally mark to show which color you got.

Put the cube back into the bag. Shake. Do this 19 more times. Make a prediction.

Do you think you will pull one color most often?

Yes   No

Cubes	Number of Times Picked	Total
red		
blue		
green		

Name ______________________________

Picture Graphs

Make a picture graph.

1. Ask 5 friends which fruit they like the best—apples, oranges, or bananas.

2. Color the fruit below.
   Then complete the graph.
   Draw and color fruit to show how many friends choose each kind.

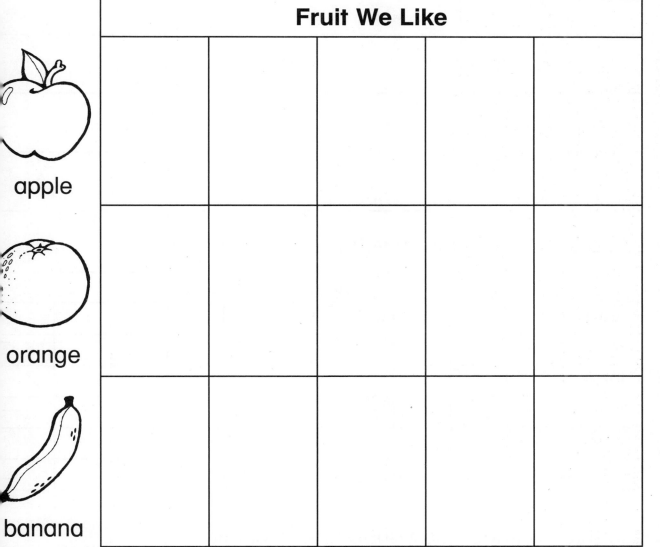

	Fruit We Like				
 apple					
 orange					
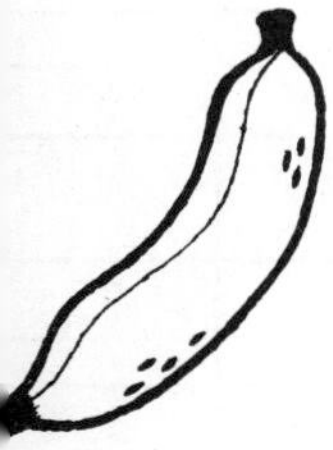 banana					

Name ______________________________

LESSON 24.2

Horizontal Bar Graphs

Roll a number cube 10 times. Make a tally mark for each time you roll the number shown on the cube. Write the total.

1.

	Tally Marks	Totals
1		
2		
3		
4		
5		
6		

2. Color the graph to match the tally marks.

	Numbers Rolled									
one										
two										
three										
four										
five										
six										

0 1 2 3 4 5 6 7 8 9

3. Which number did you roll the most times? ______

Name ______________________________________

Vertical Bar Graphs

Animals in the Park			
10			
9			
8			
7			
6			
5			
4			
3			
2			
1			
0			
	(squirrel)	(dog)	(bird)

Count the animals in the park.
Then color the graph.

Use the graph to answer the questions.

1. How many animals in all? ______
2. How many more birds than dogs? ______
3. How many fewer squirrels than birds? ______

Name ______________________________

Problem Solving • Make a Graph

15 children in the first grade wrote letters to their friends.

Noel and Kasa wrote to friends in Indiana.
April, Cassie, Chane, Drake, Kay, and Cam wrote to friends in New York.
Tena, Maria, Shawn, and Kareem wrote to friends in Florida.
Eric, Jerry, and Hope wrote to friends in Ohio.

Make tally marks to show how many letters went to each state.

Indiana ____________ New York ____________

Florida ____________ Ohio ____________

Count the tally marks. Then color the graph to match the tally marks.

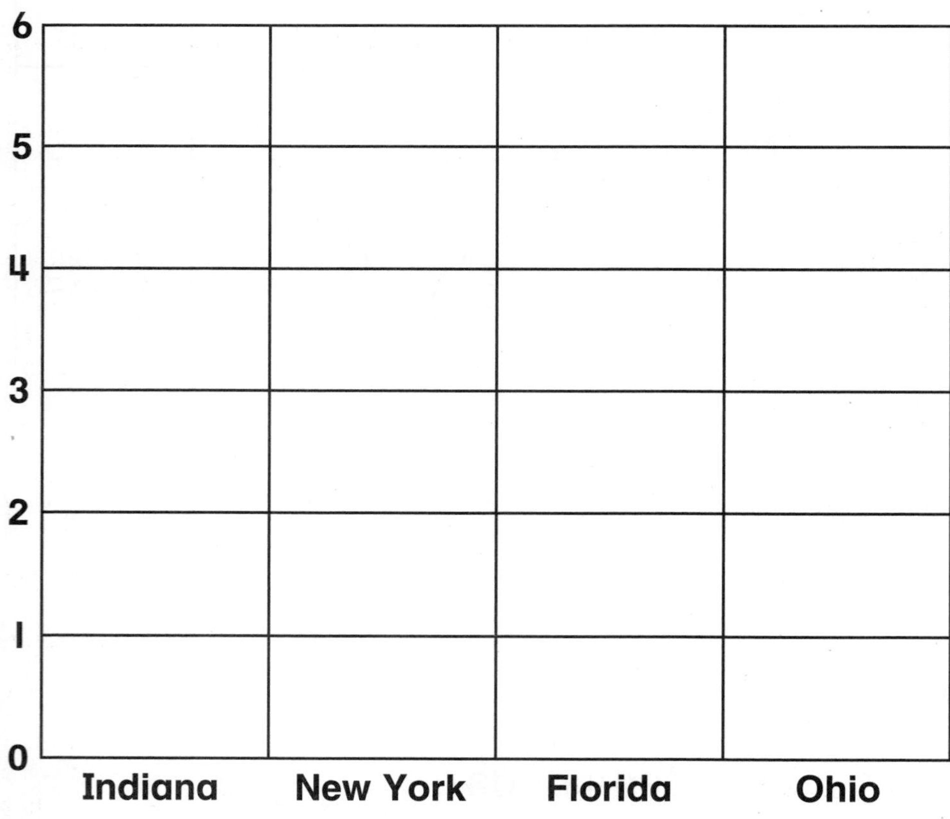

Name ______________________________

Doubles Plus One

Look at the doubles. Make doubles plus one.
Then make doubles plus two. Write the sums.

1.

6	6	6
+ 6	+ [7]	+ [8]
[12]	[13]	[14]

2.

8	8	8
+ 8	+ [ ]	+ [ ]
[ ]	[ ]	[ ]

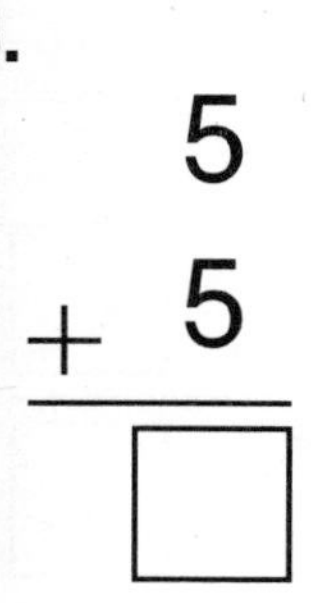

3.

5	5	5
+ 5	+ [ ]	+ [ ]
[ ]	[ ]	[ ]

4.

7	7	7
+ 7	+ [ ]	+ [ ]
[ ]	[ ]	[ ]

5.

4	4	4
+ 4	+ [ ]	+ [ ]
[ ]	[ ]	[ ]

6.

3	3	3
+ 3	+ [ ]	+ [ ]
[ ]	[ ]	[ ]

7.

2	2	2
+ 2	+ [ ]	+ [ ]
[ ]	[ ]	[ ]

8.

1	1	1
+ 1	+ [ ]	+ [ ]
[ ]	[ ]	[ ]

Name ____________________

LESSON 25.2

Doubles Minus One

Look at the doubles. Make doubles minus one. Then make doubles minus two. Write the sums.

1.
$$7 + 7 = 14 \qquad 7 + 6 = 13 \qquad 7 + 5 = 12$$

2.
$$4 + 4 = \square \qquad 4 + \square = \square \qquad 4 + \square = \square$$

3.
$$9 + 9 = \square \qquad 9 + \square = \square \qquad 9 + \square = \square$$

4.
$$3 + 3 = \square \qquad 3 + \square = \square \qquad 3 + \square = \square$$

5.
$$6 + 6 = \square \qquad 6 + \square = \square \qquad 6 + \square = \square$$

6.
$$8 + 8 = \square \qquad 8 + \square = \square \qquad 8 + \square = \square$$

7.
$$5 + 5 = \square \qquad 5 + \square = \square \qquad 5 + \square = \square$$

8.
$$2 + 2 = \square \qquad 2 + \square = \square \qquad 2 + \square = \square$$

Name ______________________________

Doubles Patterns

Joanna built sand castles.
The first day she built 1 castle.
Each day after that, she built **double** the number she built the day before, **plus 1.**
Write **C**s in the chart to show how many castles she built.

	Number of Sand Castles															
	1	2	3	4	5	6	7	8	9	10	11	12	13	14	15	16
Day 1	C															
Day 2	C	C	C													
Day 3																
Day 4																

How many sand castles did Joanna build on Day 4?

______ castles

Name ______________________________

Doubles Fact Families

Circle the numbers that make a doubles fact family. Write the facts.

1. 5 7 5 10

$$\begin{array}{r} 5 \\ +5 \\ \hline 10 \end{array} \qquad \begin{array}{r} 10 \\ -5 \\ \hline 5 \end{array}$$

2. 3 4 3 6

3. 6 9 6 12

4. 8 14 8 16

5. 7 5 7 14

6. 4 3 4 8

7. 9 8 9 18

8. 2 6 2 4

Name ____________________

LESSON 25.5

Problem Solving • Make a Model

Solve each problem.

1. Jack found 6 pails. First he found 3. How many more pails did he find?

3 pails

2. The black hen laid 4 eggs on Monday and Tuesday. On Monday she laid 2 eggs. How many eggs did she lay on Tuesday?

____ eggs

3. Little Bo-Peep lost 8 sheep. First she lost 4 sheep. How many more sheep did she lose?

____ sheep

4. The king saw 14 ships sail by. First he saw 7 ships. How many more ships did he see?

____ ships

5. Jack got 10 pails of water in all. He went up the hill 2 times. He got 5 pails the first time. How many did he get the second time?

____ pails

6. Jill got 12 pails of water in all. She went up the hill 2 times. She got 6 pails the first time. How many did she get the second time?

____ pails

Name ______________________________

Make a 10

Add in your head.
Start by making 10¢ to make a dime.

1. 9¢ + 3¢ = 1 dime and 2 pennies.
2. 6¢ + 9¢ = ____ dime and ____ pennies.
3. 8¢ + 5¢ = ____ dime and ____ pennies.
4. 4¢ + 7¢ = ____ dime and ____ penny.
5. 7¢ + 7¢ = ____ dime and ____ pennies.
6. 9¢ + 7¢ = ____ dime and ____ pennies.
7. 4¢ + 8¢ = ____ dime and ____ pennies.
8. 9¢ + 4¢ = ____ dime and ____ pennies.
9. 8¢ + 9¢ = ____ dime and ____ pennies.
10. 7¢ + 5¢ = ____ dime and ____ pennies.

Name ____________________

LESSON 26.2

Adding Three Numbers

Help each bus get to school. Add. Circle the sum that matches the number on the school.

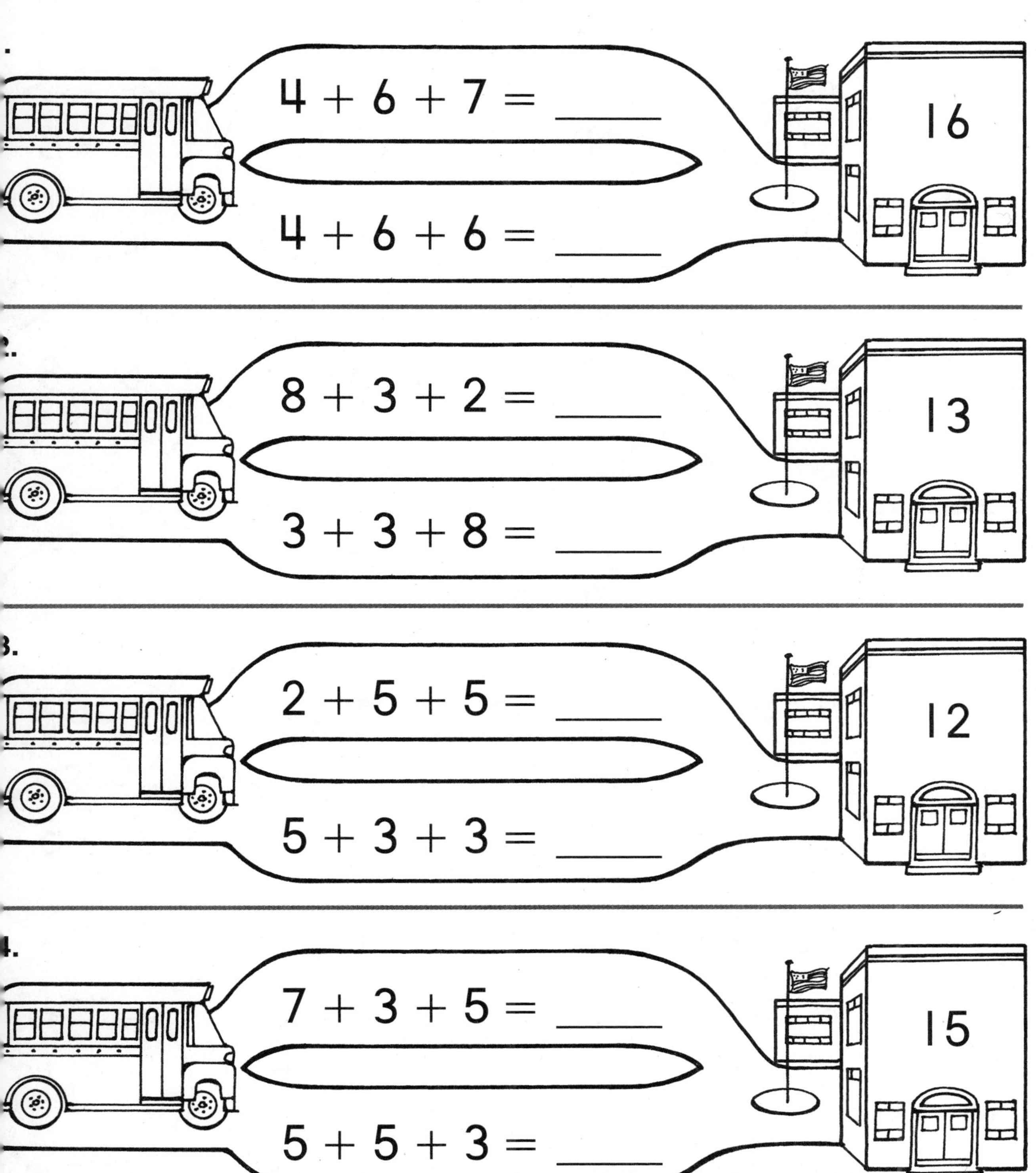

Name ____________________

Sums and Differences to 14

Find the sum or difference for each ball.
Color yellow the balls that match the number in the diamond for their row.

1.

yellow $13 - 4$ 9	$12 - 3$ ____	$5 + 4$ ____	$8 + 8$ ____

2.

$9 + 5$ ____	$7 + 7$ ____	$8 + 5$ ____	$6 + 8$ ____

3.

$9 + 3$ ____	$8 + 5$ ____	$6 + 6$ ____	$14 - 2$ ____

4.

$7 + 3$ ____	$13 - 3$ ____	$14 - 5$ ____	$5 + 5$ ____

5.

$12 - 8$ ____	$4 + 4$ ____	$13 - 5$ ____	$9 - 1$ ____

Name ______________________

LESSON 26.4

Sums and Differences to 18

Circle the names for each number.

1.

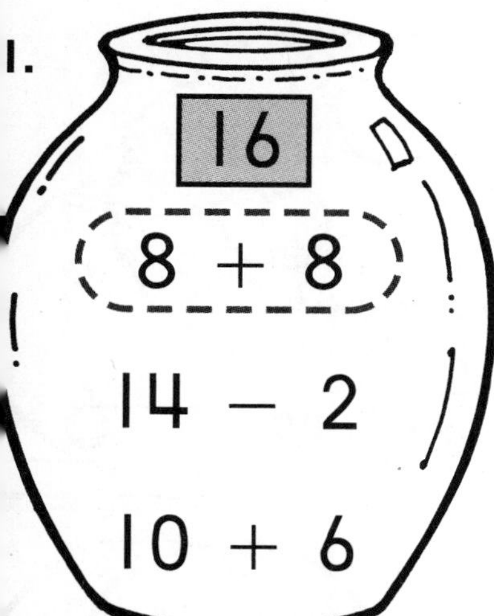

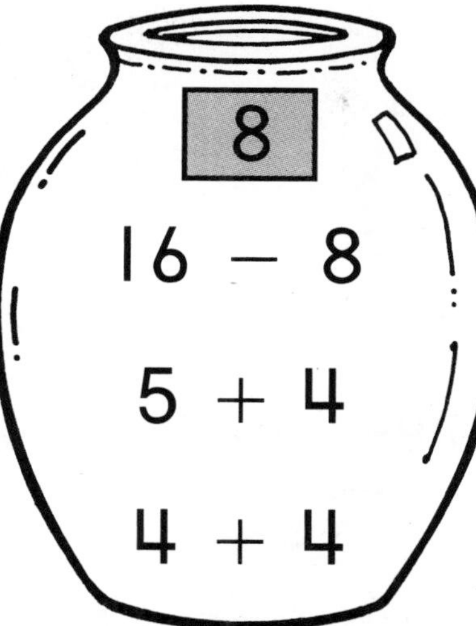

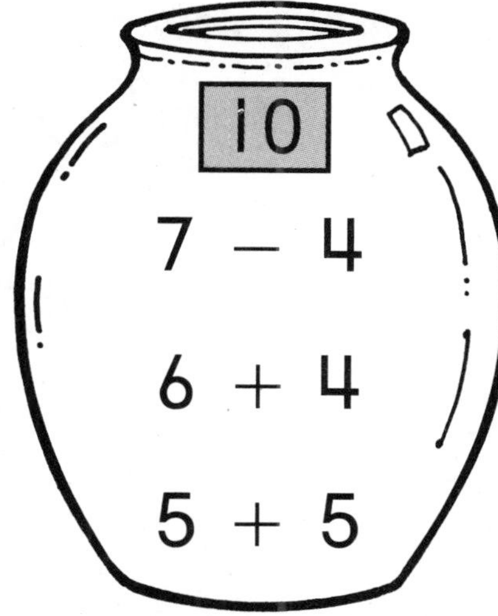

2.

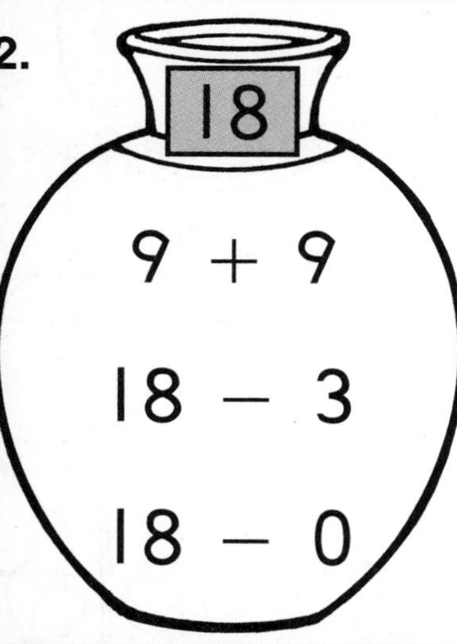

7

7 + 2

14 − 7

15 − 8

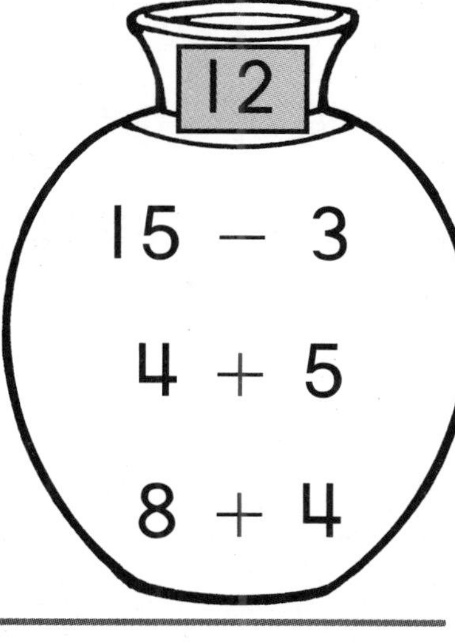

3.

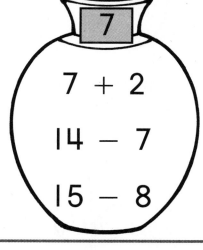

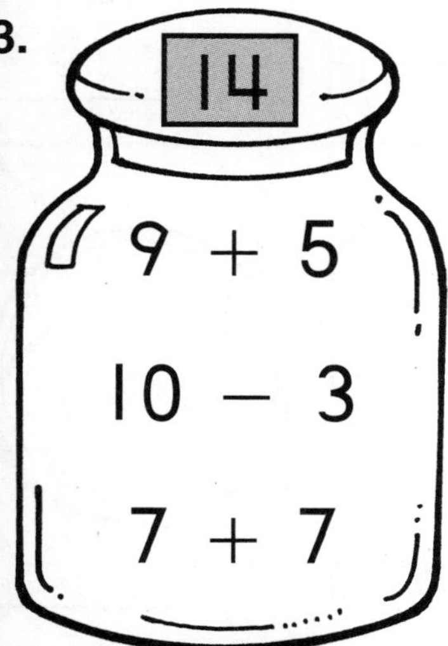

Name ______________________________

Counting Equal Groups

Draw the objects. Then write how many.

1.

Draw 3 apples in each tree. How many apples in all? 6

2.

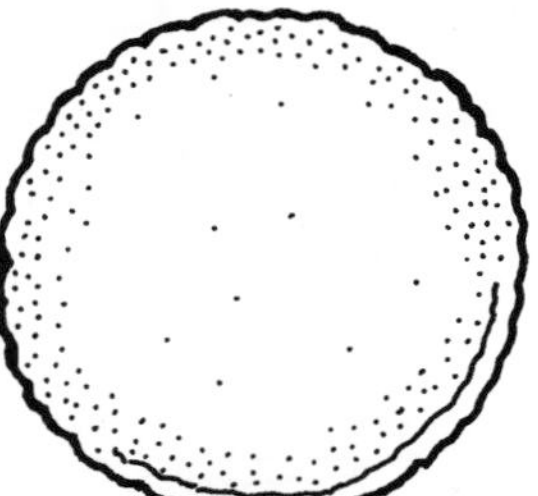

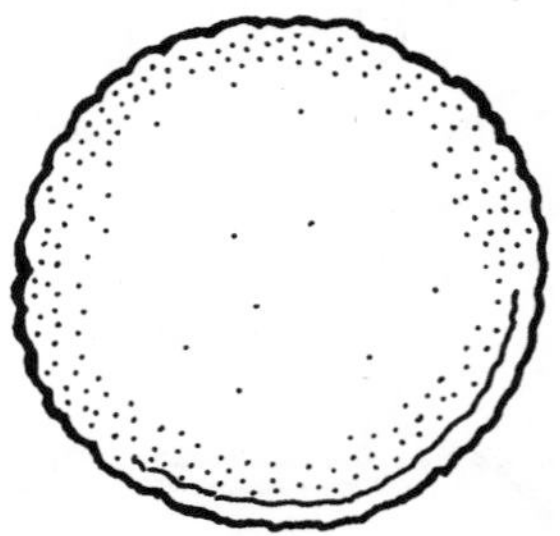

Draw 4 raisins on each cookie. How many raisins in all? ______

3.

Draw 4 petals on each flower. How many petals in all? ______

4.

Draw 3 carrots on each plate. How many carrots in all? ______

Name ______________________________

How Many in Each Group?

1. Circle 4 equal groups.

How many in each group? 2

2. Circle 2 equal groups.

How many in each group? ______

3. Circle 3 equal groups.

How many in each group? ______

4. Circle 2 equal groups.

How many in each group? ______

5. Circle 3 equal groups.

How many in each group? ______

Name ____________________

How Many Groups?

1.

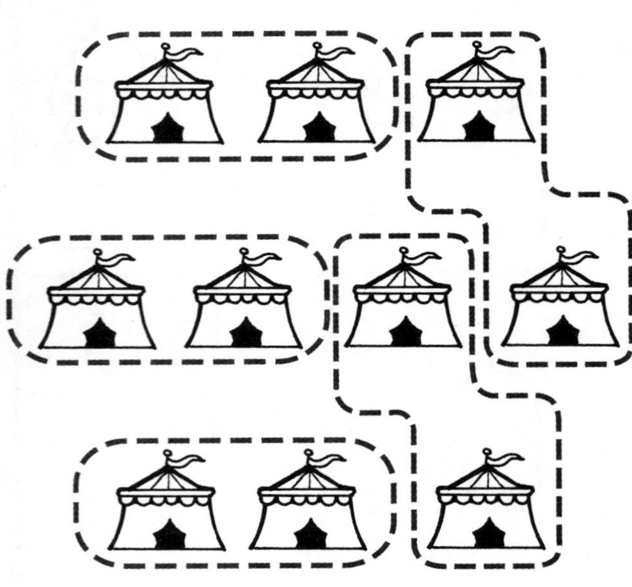

Circle groups of 2.
How many groups? 5

2.

Circle groups of 4.
How many groups? ______

3.

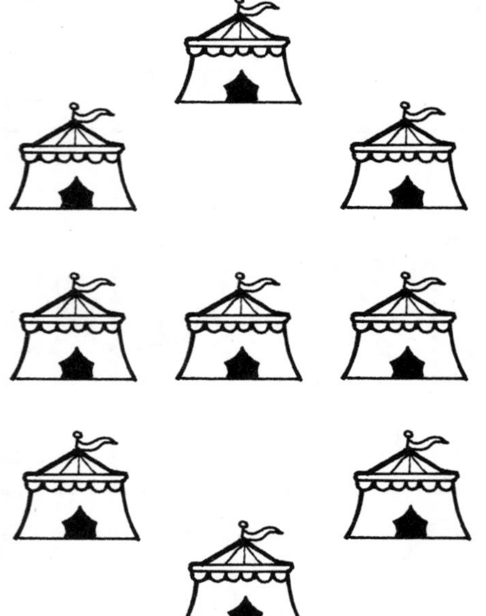

Circle groups of 3.
How many groups? ______

4.

Circle groups of 5.
How many groups? ______

Name ____________________

LESSON 27.4

Problem Solving • Draw a Picture

Fill in a number. Draw a picture to solve.

1. There are 3 clowns.

   Each clown has ____ balloons.

   How many balloons are there in all?

   ____ balloons

2. There are 6 stuffed animals.

   Each child has ____ animals.

   How many children will get animals?

   ____ children

3. There are 9 boxes of juice.

   We drink ____ of the boxes.

   How many boxes of juice are left?

   ____ boxes

4. There are 3 rows of flowers.

   Each row has ____ flowers in it.

   How many flowers are there?

   ____ flowers

Name ______________________________

LESSON 28.1

Adding and Subtracting Tens

Write how many you add or subtract.
Color each flower to match its number.

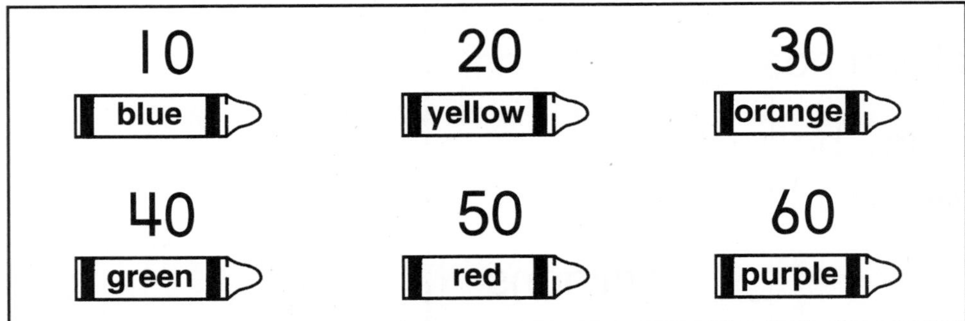

1. 20

40

2. 30

− ◯

20

3. 40

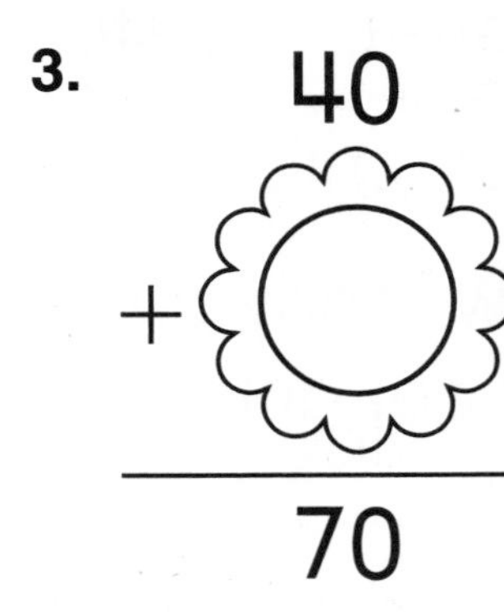

70

4. 10

+ ◯

60

5. 20

+ ◯

80

6. 50

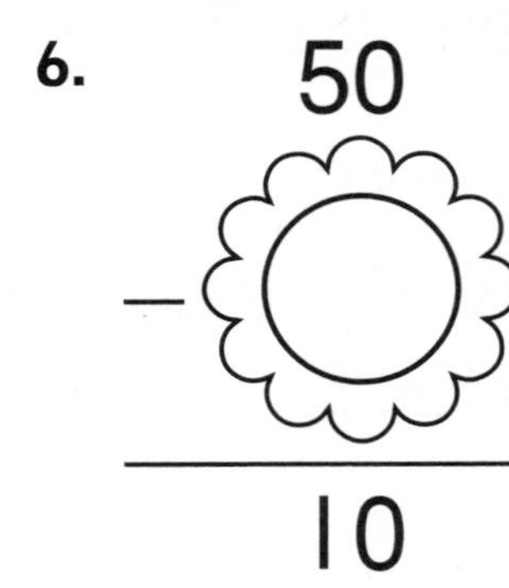

10

7. 40

− ◯

10

8. 70

− ◯

50

9. 80

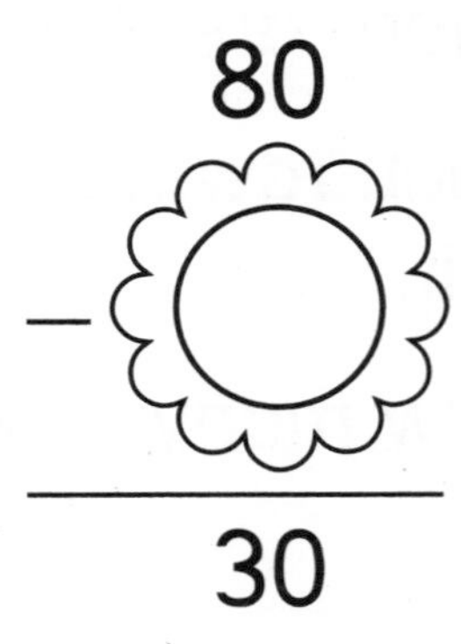

30

Name ____________________

LESSON 28.2

Adding Tens and Ones

The tables show how many stickers each child got each month. How many stickers did each child get in all?

April	
Name	**Stickers**
Jack	24
Maria	21
Cole	20
Etta	18
Tania	12
Lindsay	22

May	
Name	**Stickers**
Jack	30
Maria	22
Cole	18
Etta	11
Tania	14
Lindsay	27

1. Jack

	2	4
+	3	0
	5	4

2. Maria

+		

3. Cole

+		

4. Etta

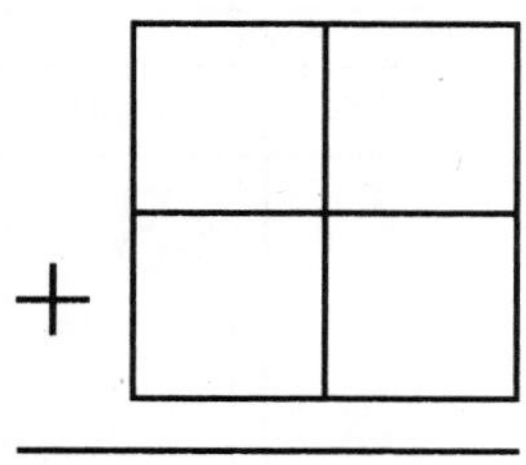

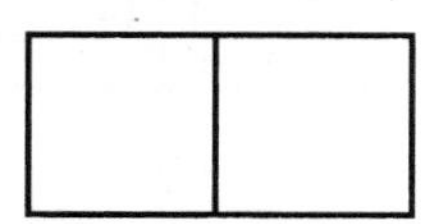

5. Tania

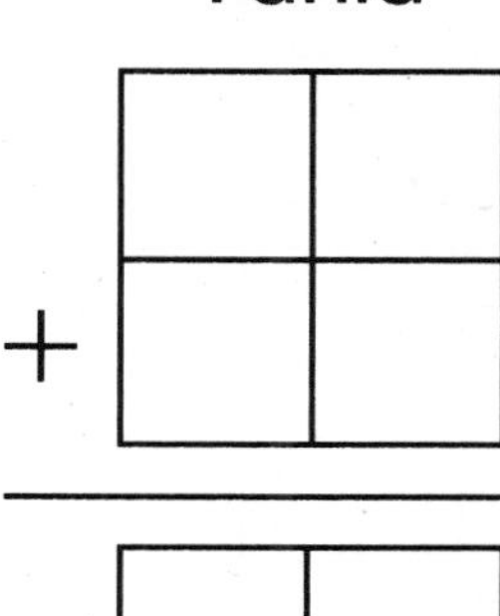

6. Lindsay

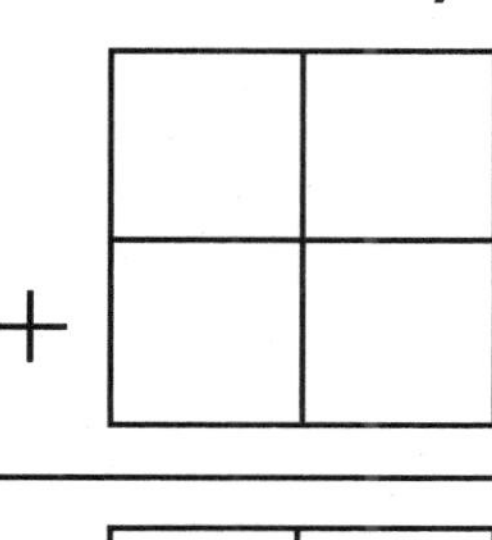

Name ____________________

Subtracting Tens and Ones

Use a pencil and a paper clip to make a spinner.
Spin the paper clip two times for each problem. Write each number in a box.
Subtract.
Write how many are left.
Use punch-out tens and ones if you need to.

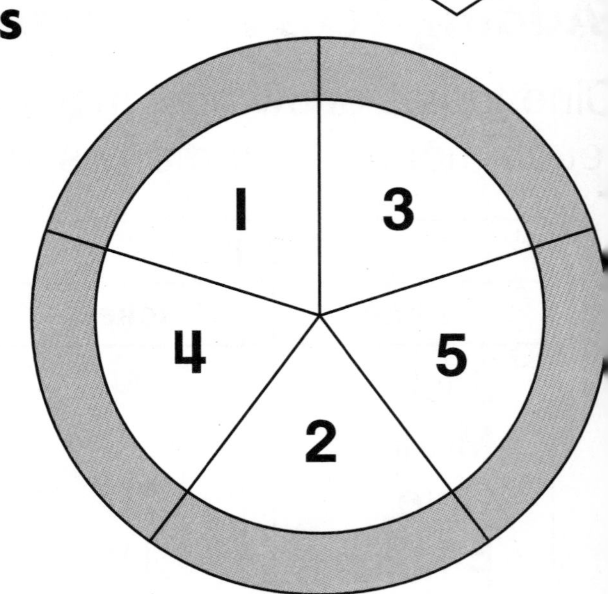

1.

tens	ones
6	5
− 4	3
2	2

2.

tens	ones
8	9
− □	□

3.

tens	ones
9	7
− □	□

4.

tens	ones
7	6
− □	□

5.

tens	ones
8	5
− □	□

6.

tens	ones
6	8
− □	□

7.

tens	ones
6	6
− □	□

8.

tens	ones
5	5
− □	□

9.

tens	ones
7	8
− □	□

Name ______________________

Reasonable Answer

Circle the answer that makes sense.

1. Mark has 47¢.
He buys a ball for 20¢.
He buys a balloon for 7¢.
How much money does he have left?

50¢

47¢

20¢

2. Keesha has 11 colored pencils. She buys 7 more on Monday. She buys 1 more on Friday. How many does she have now?

11 pencils

19 pencils

7 pencils

3. Suki has 39¢. She spends 11¢ for a sticker. She spends 22¢ for a notebook. How much money does she have left?

6¢

39¢

69¢

4. Carlos drew 51 blue stars. Then he drew 31 purple stars. The next day he drew 11 green stars. How many stars did he draw in all?

11 stars

93 stars

51 stars